Bugwatching

Bugwatching

The Art, Joy, and Importance of Observing Insects

Eric R. Eaton

Illustrated by Samantha Gallagher

Princeton University Press

Princeton and Oxford

Published by Princeton University Press
41 William Street, Princeton, New Jersey 08540
99 Banbury Road, Oxford OX2 6JX
press.princeton.edu

GPSR Authorized Representative: Easy Access System Europe -
Mustamäe tee 50, 10621 Tallinn, Estonia,
gpsr.requests@easproject.com

All Rights Reserved
ISBN (pbk.) 978-0-691-26400-4
ISBN (e-book) 978-0-691-26529-2

British Library Cataloging-in-Publication Data is available

Editorial: Robert Kirk and Megan Mendonça
Production Editorial: Karen Carter
Text Design: D & N Publishing, Wiltshire, UK
Cover Design: Ben Higgins
Production: Steven Sears
Publicity: Matthew Taylor and Caitlyn Robson-Iszatt
Copyeditor: Patricia Fogarty

Cover images from iStock

This book has been composed in Garamond Premier Pro

Printed in China
10 9 8 7 6 5 4 3 2 1

CONTENTS

To STEVEN PRCHAL May 20, 1950–April 17, 2015, who coined the word *bugwatching* and was a mentor to so many; and to ERIN STARKEY March 21, 2003– September 5, 2019, whose light shined entirely too briefly.

(Photographer unknown)

(Christina Starkey)

PREFACE

ONE OF MY vivid childhood memories is that of an insect. Two, actually. I heard a male Fork-tailed Bush Katydid, *Scudderia furcata*, calling from a rhododendron in front of our Portland, Oregon home. Shortly after I spotted the lovely green insect, a female flew down to join him. They soon began copulating, about the time my father arrived home from work. "Dad, come look at the katydids mating!" I breathlessly exclaimed. Eventually, a gelatinous white mass appeared where the two katydids were joined. I thought something had gone horribly wrong. Decades later, I learned male katydids, and some other members of the order Orthoptera, transfer sperm in a protein-rich package called a spermatophore. Both katydids were healthy after all.

There are few resources to help the amateur insect observer comprehend such behaviors or learn how to do so. "Field guide companion" books help birders become more effective at observing and recording our avian friends, but there has not been a comparable reference for bugwatchers. The late Steven J. Prchal probably coined the word *bugwatching*, but his short-lived quarterly journal, *Backyard Bugwatching*, never became as popular as, say, *Birdwatcher's Digest*.

For the sake of simplicity and familiarity, the word "bug" is used in this book to mean any kind of insect. Technically, only the species in the insect order Hemiptera qualify as "true bugs."

A Fork-tailed Bush Katydid, the author's "spark bug" that ignited a lifelong interest in insects and their kin. (ERIC R. EATON)

— ABOUT THIS BOOK

This reference seeks to achieve three goals. The first two: Promote bug-watching as a viable recreational pursuit and help solve the challenges of watching insects. How does one observe insects safely and ethically? How does one develop a search image for insects or know where to look for certain species? What insect behaviors are you likely to see? Why are insects so difficult to identify? Where does one find additional resources and connect with other bugwatchers? Why is it even important to be a bugwatcher? This book will answer those questions.

The third mission of this book is to expose the assumptions we make about other people who share an interest in bugwatching or natural history in general. Most of us, the author included, are guilty of ableism, gatekeeping, and other discriminatory practices, even if by accident. We need to train ourselves to be more aware of differences and make appropriate accommodations. The title of this book is not meant to suggest that the only way to understand and appreciate insects is through vision, nor does it endorse the surveillance culture we live in. Profiles of inspirational individuals are dispersed throughout the book, demonstrating the human diversity of the bugwatching community and perhaps serving as a mirror to your own experience.

No single volume, nor even an encyclopedia, can cover every aspect of the natural history of every insect. Professional entomologists specialize in a particular family, or even genus or species. Understandably, they often focus on economically important pests. That means we know collectively little about most other insects. With so much yet to discover, we need as many bugwatchers as possible to fill in the gaps (chasms, really).

It is the author's wish that you will not read this book from cover to cover, but be inspired to put it down frequently and go outside to begin a lifelong quest for insects. You need not even go outdoors. Many species live with you in your home, up in the attic, out in the garage, or in the basement. Your odds of observing something new in the insect realm are vastly greater than winning the lottery, and arguably more rewarding.

— A BRIEF HISTORY OF ENTOMOLOGY AND BUGWATCHING

The study of insects has been driven by two forces. One of those is personal safety and, by extension, the welfare of our families, pets, livestock, crops, forests, and other collective resources. This falls under the umbrella of economic entomology, focused primarily on pests. The flipside of the entomology coin is sheer curiosity and the desire to create order out of perceived chaos.

What we know of insects that are not economically important is largely due to people who were compelled to gain insights into a given species, regardless of available funding and other resources. This scenario frequently applies to ecological, evolutionary, and field entomology, all of which are aspects of bugwatching, albeit mostly in the academic sense.

As an example of the influence of casual observation of insects, much of our early knowledge of the life histories of aquatic insects came from fly-fishing. In England, the hobby dates back to at least 1450, as revealed by an early manuscript from Dame Juliana Berners that was published in 1496 as *The Treatise of Fishing with an Angle*. Izaak Walton's *The Compleat Angler* is better known, originally published in 1653 and still in print.

The dawn of microscopy in the 17th century opened a whole new world. The smallest of creatures could no longer be dismissed as simple and unworthy of admiration. We were confronted with details that demanded we look beyond the familiar mammals and birds to the insects, with or without the aid of magnification.

The classification of organisms dates back to Aristotle's crude hierarchy, atop which he placed our own species. Formal taxonomy began when René Antoine Ferchault de Réaumur, a French mathematician and naturalist, published a systematic treatise on insects in the 1730s. Immediately thereafter, biologist Carl Linnaeus created the binomial system of scientific names (genus and species), in Latin and Latinized Greek. This is the system we still use.

Among early modern entomologists, there were celebrated bugwatchers. Jean-Henri Fabre observed insects around his home and in nearby landscapes in France from the 1870s to the early 1900s. He took copious notes, which he crafted into a ten-part series, *Souvenirs entomologiques*. These were captivating stories—some would call them anthropomorphic—and his works have been widely translated. They remain classics of natural history literature and are immensely popular in Japan.

Women made profound impacts, too, but have been historically overlooked, their contributions minimized. Mary Treat was most active in the same period as Fabre, but in New Jersey, USA. Treat was a well-rounded naturalist who could write and speak to scientific professionals and the lay public with equal command. One of her books was *Injurious Insects of the Farm and Field*, first published in 1882 and reprinted five times. *Home Studies in Nature* (1885) was more general, the output of observations made close to her home.

Fast forward to the 20th century. The Amateur Entomologists' Society was founded in the United Kingdom, by Leonard Tesch, in 1935. It was initially known as the Entomological Exchange and Correspondence Club; the name change happened in 1937. It took until 1972 for membership to exceed one thousand, and the society became a registered charity (a type of nonprofit) three years later. In 2005, the AES established formal relations with the Royal Entomological Society, an organization of professional entomologists. The partnership continues to grow, allowing bugwatchers to partner with academic scholars in research and conservation endeavors.

In the United States, the American Entomological Society was founded in 1859 in Philadelphia. Not to be confused with the Entomological Society of America (ESA), the AES explicitly welcomes all who are interested in entomology, whether amateurs or professionals.

Colleen Seeley founded the Teen International Entomology Group (TIEG) in 1965, with the help of both 4-H and the Cooperative Extension Service at Cornell University in Ithaca, New York. The organization published a newsletter that eventually metamorphosed into a quarterly journal of high quality, with regional editors in the United States and Canada. Many members were gifted writers, illustrators, or photographers. Advertisements in the newsletter encouraged correspondence among members, trading of specimens, and scheduling of field outings.

TIEG began to decline in membership and activity in the late 1970s. It was formally resurrected as the Young Entomologists' Society (YES), by Gary A. Dunn in 1984. Based in Lansing, Michigan, the headquarters featured the Minibeast Zooseum and Education Center, which provided school tours and outreach programs under Dunn's energetic leadership. YES boasted 7,000 members at its peak, produced three periodicals, and, at one point, maintained ten websites. Sadly, the organization folded in 2010.

Community science can be viewed as bugwatching for the digital age. Observations occur in real life, but the sharing and storytelling is usually online. It can be a solitary or social pursuit, and as general or specific as one chooses. Community science projects are constantly created, and they are not all related to entomology, naturally.

Have we come a long way since the Victorian era, when having a cabinet of curiosities was a status symbol for wealthy elites who procured specimens from faraway lands through the exploitation and servitude of Indigenous peoples? Perhaps, but there is still a substantial degree of White advantage in access to the tools needed for nature study and access to places in which to do it.

CHAPTER 1

THE REWARDS OF BUGWATCHING

What sort of insects do you rejoice in, where you come from?
———— THE GNAT, IN *THROUGH THE LOOKING GLASS*, BY LEWIS CARROLL

BIRDERS CLAIM THEIR pursuit is accessible to anyone because birds are common and easily spotted; but if birds are everywhere, insects are everywhere-er-er. Insects are wildlife, too, and provide much of the same satisfaction for the naturalist as birds do, but in greater quantity and diversity.

Like any outdoor activity, bugwatching usually provides some degree of exercise and exposure to a dose of vitamin D (the sun), and it generally improves one's mood. It can take you to extraordinarily unique and scenic landscapes. For some, nature recreation is also a spiritual experience, a way of reconnecting with other species. Oh, the stories you will have to tell.

— VISUAL BEAUTY

FEW, IF ANY, members of the animal kingdom exceed insects in their rainbow of dazzling colors, grace in movement, and vibrancy of life. Large groups

Gorgeous: Insects exhibit vibrant beauty, in all the colors of the rainbow, sometimes all on one insect, like this Emerald Flower Scarab, *Euphoria fulgida*, from Colorado, USA. (ERIC R. EATON)

of insects are a spectacular reminder of the exuberance of living organisms, the indulgence of the Creator, or, conversely, the marvels of evolutionary mechanisms. Awe, wonder, curiosity, respect, and, ultimately, affection are all spawned in whole or in part by one's perception of beauty.

There is little argument that butterflies and dragonflies are extraordinarily beautiful, so much so that they are often considered honorary birds. We see ladybird beetles as cute and colorful, and since they are small, they evoke no feelings of intimidation. Fireflies are enchanting because we see their flashing light organs, not the beetles themselves. It is easy to imagine them as faeries come to life.

It takes a more learned eye, perhaps, to find appeal in other insects. Maybe it simply takes a closer look. There are many harmless beetles and true bugs that defy artistic imaginations of color combinations and patterns. Many a wasp has spectacularly iridescent, shimmering wings.

Insect beauty can come in the most unexpected places. The eyes of many flies, especially horse flies and deer flies in the family Tabanidae, can be striped or spotted with green, red, blue, or gold. Allow yourself to be mes-mer-EYES-ed by them at your own risk. Female tabanids can deliver a painful bite. At night, the eyes of moths may glow red in the beam of your flashlight.

— SONG

INSECTS PROVIDE FOR a multisensory experience, including "vocal" performances. While it might be a stretch to equate the monotonous, blaring noise of a male cicada to the melodious twittering of a handsome warbler, the sounds serve the same function: they attract potential mates. We probably can agree that the hours after dark are enhanced by the serenades of crickets and katydids.

— SPECIES DIVERSITY

INSECT DIVERSITY IS seemingly inexhaustible. Even a single species may have several color forms, exhibit sexual dimorphism, or otherwise be variable. Complete metamorphosis means that one life stage differs radically from another. It is impossible to become bored if one chooses to investigate insects.

Leafhoppers, mostly small, true bugs in the family Cicadellidae, are a case in point. There are approximately 3,000 species in North America north of

OPPOSITE: Melodious: Male tree crickets, *Oecanthus* spp., produce a soothing song by rasping their front wings together. (SAMANTHA GALLAGHER)

Staggering diversity: Leafhoppers, true bugs in the family Cicadellidae, are among the more species-rich of all groups of insects. Most are small, 5 millimeters or less in length. (SAMANTHA GALLAGHER)

Mexico, with about 22,000 species known globally. That pales in comparison to rove beetles in the family Staphylinidae. There are at least 64,000 species worldwide. Certain kinds of flies and wasps may prove to be even more diverse. We know that little of the natural world.

Pure joy: a group of children in Madagascar delight in reconnecting with insects like this grasshopper, long confined within the borders of a nearby national park, where steep permit fees have severed local ties to flora and fauna. (ALLISON F. MONROE)

— PERSONAL DISCOVERY

ESPECIALLY FOR THE beginner, the potential for new revelations about insects abounds every time one goes outdoors. Insects are the perfect vehicles to ignite curiosity and a sense of wonder in children of all ages. Be it deciphering the nighttime symphony of katydids and crickets, breaking the code of the flashing of fireflies, examining the texture of a June beetle, or comparing the odors of various stink bugs, you can achieve great personal satisfaction. Observing the metamorphosis of a butterfly or the flight patterns of dragonflies can become a healthy, all-consuming passion.

— INSPIRATION

INSECTS CAN BE a source of inspiration for art, and they foster intangible qualities like persistence or patience. The butterfly that is barely slowed down by tattered wings symbolizes durability and resilience when we face our own personal obstacles, be they physical or psychological.

Science and engineering are replete with examples of invention originating from observations of insects. The origin of papermaking is frequently attributed to China's Cai Lun, during the Han Dynasty, in about 100 CE, as a result of his observation of the work of paper wasps. Centuries later, when shortages of materials like cotton and linen created hardship for the paper-making industry, wasps came to the rescue again. In 1719, René Antoine Ferchault de Réaumer, a French naturalist and physicist, noted paper wasps using wood fibers and their saliva to manufacture pulp used to build their nests. This led to new means of the industrial production of paper still used today.

Chainsaw design was improved when Joe Cox observed the feeding behavior of larvae of the Ponderous Borer, *Trichocnemis spiculatus*, tunneling through wood with an enviable degree of efficiency. The opposing mandibles of the insect inspired Cox to create a chain of alternating "right" and "left" teeth that cut in a similar manner. Cox received a patent and began production from his basement in 1947. Oregon Sawchain Corporation later became Omark Industries, now Oregon Tool, Inc.

"Butterfly"
"Papillon"
"Mariposa"

"Moth"
"Papillon de nuit"
"Mariposa nocturna"

In any language: "Butterfly" and "moth" in English, French, Spanish, and American Sign Language. Accommodating non-English speakers expands our own vocabulary. (SAMANTHA GALLAGHER)

Handicap parking with bumble bee.
(ANNE READEL)

The impact of insects continues to this day. The architecture of the Exoskeleton Tower in Cheongna City, South Korea, was inspired by termite mounds. A new surgical tool was inspired by the flexible ovipositor of a wasp.

Insects are also metaphors for transformation. Insect metamorphosis has long been seen as a magical phenomenon. The cellular clusters within the larval stage that are destined to become features of the adult insect are called, appropriately, imaginal discs. The idea that we can imagine ourselves as greater, stronger, and better can be easily prompted by observing the growth of insects.

— MAKING CONTRIBUTIONS TO SCIENCE

IN AUGUST 2021, the discovery of two colonies of stingless bees, *Plebeia emerina*, in Palo Alto, California, by a four-year-old girl made national news. Stingless bees (family Apidae, tribe Meliponini) are tropical and subtropical, so what were they doing that far north? It turns out that the bees had been sent to Palo Alto over 70 years ago. A researcher at Stanford University received ten colonies from a colleague in Brazil in 1948. In 1950, other colonies were sent to federal bee research labs in Arizona, Utah, Maryland, and Louisiana. Only the Palo Alto colonies lived more than a year. Thus ended the potential of the foreign bees to be supplemental pollinators of American crops. Stanford professor Georg Schafer kept one colony in his backyard, but after he passed away in 1962, it was assumed his stingless bees did, too. It was not until 2013, when a homeowner near Schafer's old residence found a colony in a tree that the bees came back on the radar.

You, too, can make groundbreaking discoveries. You may be the first to detect a species in your county or state or aid in tracking changes in distributions as a result of climate change. You might record a new host plant or animal for an insect. You could discover a species new to science. The possibilities are endless.

Happiness is: A trio of volunteers release captive-reared Palos Verdes Blue butterflies into the wild. Participating in such conservation endeavors is highly rewarding. (MAX A. SPRUTE)

— PARTICIPATION IN CONSERVATION OF BIODIVERSITY

EVEN BEFORE HEADLINES about an "insect apocalypse," many people became interested in the conservation of insects like the Monarch butterfly. That interest and concern has now expanded to include the welfare of other insects, especially pollinators like bees. Bugwatching skills are essential in order to make conservation efforts effective. Community science projects abound, and you can initiate your own as well, to help others learn how to observe organisms in an accurate and strategic manner that meshes with the efforts of professional entomologists and ecologists.

— SOCIAL INTERACTION

BUGWATCHING WAS ONCE a solitary pursuit. That was before the internet made everything cool. Today, it is easier to find like-minded souls eager to share their enthusiasm, observations, and experience. Professional entomologists are more accessible thanks to social media, email, and podcasts. There are

Adventuring together: Volunteers enjoy the great outdoors while they survey for butterflies in Washington state. (KAREN POVEY)

annual and one-time events, community science projects, and online interest groups. There are virtual events in real time through Facebook, YouTube, and other channels. See chapter 7 for more about social bugwatching.

— SHELLY COX (SHE/HER)

HI! I CAN thank my grandfather for fostering my interest in nature. He had a childlike love of nature himself, and even though he lived in the heart of a city, he always managed to find wildlife of some kind. He bought me my first field guides, which I still own today, and took me outside to investigate an interesting bird, bug, or other creature. One time, he found a large spider in the basement. Having no idea what it was,

Shelly Cox (SHELLY COX)

we made a trip to the local college, where a professor identified it as a wolf spider. Reassured that it was not dangerous and instead a beneficial predator of pests, we released it back into the basement. My grandfather told me it would be best not to tell grandma.

Today, I am fortunate to live on 86 acres in a rural area, with additional family-owned farmland relatively close by. I spend a lot of time walking the properties, seeking new-to-me species to photograph. I also do nature journaling and illustration. My family and I also enjoy visiting state parks with a variety of different habitats, and hike, camp, and kayak.

Here in northwestern Missouri, one major challenge to bugwatching is the scale of agriculture. Farmers are beginning to plow roadway to roadway and fencerow to fencerow, eliminating edge habitats where I would typically find a variety of species to document. I am trying to help overcome this by educating people in our region about the importance of maintaining even the smallest parcels of habitat for our declining populations of beneficial insects.

By profession, I am employed as a naturalist for the Remington Nature Center in St. Joseph, Missouri, where I interpret the natural world for children and adults. My greatest satisfaction comes from watching a child's fear or uncertainty turn to fascination. I author an article in the local newspaper each week, to cultivate a better public understanding of local wildlife, especially insects. I also founded Tortoise Run Farm, a rescue for tortoises and box turtles.

It is sometimes a challenge for people, like me, who lack an advanced degree to gain a level of respect in the sciences. We have to carve a nontraditional path. I have written two "bug" books for children, and coauthored another on amphibians and reptiles. To have people recognize me as the "Bug Lady" makes me smile, and I appreciate being considered an expert.

My advice to new bugwatchers would be to invest in some good field guides, learn to take notes and/or journal, and reach out to others who share your interest. Many experts are happy to take you under their wing and teach you all they know. Also, participating in community science projects is an exciting and rewarding way of learning and networking.

SHELLY'S BLOG: https://mobugs.blogspot.com/

CHAPTER 2

HOW TO WATCH BUGS

Walk slow. Look close. Be curious.

MJ HATFIELD

INSECT OBSERVATION IS most rewarding when waiting, rather than pursuing. Selecting a small area, easily scanned, with your back to the sun, is ideal. Forest edges, small ponds or even puddles, and patches of wildflowers are good places to start. Your own yard or garden may easily suffice.

If you want to walk or hike while looking for insects, Chris Kline of Butterfly Ridge and the Polygonia Foundation offers these tips:

— TAKE IT SLOW. Insects can detect motion much better than most other animals. You can also easily overlook camouflaged and perched insects if you go at a brisk pace.
— RETRACE YOUR STEPS. Consider returning along the same trail or repeating your trek several times. Insect activity changes frequently, and often drastically, over short periods.

Think small: Adjust your search image to macro mode, and insects will soon materialize. A hairstreak butterfly at Lady Bird Johnson Wildflower Center, Texas, USA. (MIKE QUINN)

— WATCH FOR SHADOWS of flying insects, especially large butterflies, and dragonflies. Use your peripheral vision to detect flying insects, too.
— PAUSE FREQUENTLY. Benches along trails and in public parks and gardens are not only for resting. They are stations to make observations from.
— TO GET CLOSER to an insect, move slowly and avoid casting a shadow over it.

Your experience can be enhanced by taking precautions before you leave home and by purchasing or borrowing equipment and references that aid in observation and identification of what you see. Recording your observations increases their value to others, while making memories for yourself. Don't forget to use senses other than vision, but have reverence and respect for insects when you handle them.

— SAFETY FIRST

IN OUR EAGERNESS to get outside, we frequently ignore necessities that make nature exploration more comfortable, if not survivable. Take the following precautions:

— Apply sunscreen, and pack the product for reapplication later.
— Carry more water than you think you will need.
— Include food, or at least snacks, even if you are not watching your blood sugar.
— Carry a basic first-aid kit.
— Pack extra batteries for GPS, flashlights, and other devices. Charge your phone ahead of time.
— Wear a hat, proper clothing (long sleeves, pants), and durable footwear. A bright vest is a good idea during hunting season. Official-looking gear, even a T-shirt with a logo, can evoke a positive, or less suspicious, reaction from other people.
— Consider bear spray, mace, or other self-defense products in the event of encounters with ill-intentioned humans, as well as dangerous wildlife.
  A whistle may be helpful, too.
— Go afield with at least one other person, especially if going to remote or unfamiliar locations.
— Let other parties know where you are going, when, and about how long you will be there.

— Plan escape routes, and leave an area if you are feeling uneasy.
  Trust your instincts.
— Pause frequently to hydrate.

Insect repellents sound counterintuitive, but such products can increase your comfort level, especially in wetlands where mosquitoes abound. They also decrease the likelihood of acquiring viruses and other arthropod-borne illnesses. Repellents will have no effect on anything but biting flies and ticks. Be careful when handling dead insects. They may have been poisoned, so wash your hands afterward.

Before entering private property or restricted areas on public lands, secure written permission to be in that location at an agreed-upon date and time. This is more than common courtesy; it is a means of finding you if you become lost, and it establishes legal foundations for rights and liability if you have an accident. Be an exemplary guest, always.

— FIRE ANTS

THE RED IMPORTED Fire Ant (RIFA), *Solenopsis invicta*, poses a serious hazard in places where it is native or introduced, including much of the southern United States. An accidental encounter can ruin your day, and the effects of stings can persist for weeks. These ants construct mounds of variable dimension, but the ants are seldom seen on the surface. It is easy to step in a mound without recognizing it, and the consequences can be exceptionally painful. Further, RIFA can also nest under boards, logs, and other objects, so the act of lifting an object to look for other insects can result in ants swarming over your hand immediately.

— OTHER DANGEROUS INSECTS

ASIDE FROM THE obvious stinging wasps, bees, and ants, there are other insects to be aware of. Chief among these are caterpillars with stinging spines. The larvae of some giant silkmoths, namely the Io Moth, *Automeris io*, and buck moth caterpillars, genus *Hemileuca*, are examples. Slug caterpillars in the family Limacodidae, including the Saddleback Caterpillar, Spiny Oak Slug, Spun Glass Slug, and Stinging Rose Caterpillar, are relatively small, but pack a powerful sting. The caterpillars of flannel moths, family Megalopygidae, are deceptively furry. Beneath the long, soft, silky hair are stinging spines.

Hairy caterpillars that cannot sting may still cause skin irritations such as contact dermatitis, so think twice before picking up a tussock moth caterpillar or a tiger moth larva.

Danger! A Stinging Rose Caterpillar, *Parasa indetermina*, looks like ribbon candy, but is studded in venomous spines. A caterpillar of an Io Moth, *Automeris io*, is similarly armored. (ERIC R. EATON AND K. LEEKER, RESPECTIVELY)

Blister beetles in the family Meloidae are locally common; dozens or hundreds of individuals are often present in a small area. Squeeze one at your peril. They are soft-bodied, but when molested, exude hemolymph (insect blood) from their body joints. The viscous fluid contains cantharidin, a highly irritating chemical that can raise scarring blisters on sensitive skin. So potent is cantharidin that livestock can die from consuming hay that has been bailed with blister beetles.

Few insects bite, even when handled, but it is best to avoid most assassin bugs, family Reduviidae, and aquatic true bugs like backswimmers and giant water bugs. They all inject venom and/or digestive enzymes, designed to kill prey and begin the digestive process. When unleashed for self-defense, the results are excruciating for the victim. The bites cause pain, but other symptoms vary with individual human immune system responses.

— RELATIVE COMFORT AND ACCOMMODATING DISABILITIES

I like to sit (well, crouch) in one spot for a good while and let micro movements catch my eager eyes. Lots to observe!

— SHANNON BOWLEY

PERHAPS YOU HAVE a knee injury, your lower back is prone to chronic soreness, or you are afflicted by other maladies. Taking along a stool or cushion(s) may improve your ability to sit for a longer spell than usual. Kneepads allow for getting lower still, permitting an eye-to-eye perspective with insects in tall grass and on flowers, the surface of a pond, and the surface of the ground itself. A yoga mat is perfect if you want to lie prone in relative comfort. An umbrella or parasol can afford shade.

The best parks, preserves, and trail systems allow for access by wheelchairs and recognize the need for frequent stops, with numerous benches along paved or smooth substrates or boardwalks. Signage, if present, is easily read from any height. Unfortunately, special trails and other features for the visually impaired are still sorely lacking in most natural parks. Please rate parks after your visit, with special attention to their accommodations of the disabled. Encourage all parks to make improvements in this regard. If you are an able-bodied insect-watcher, consider offering to guide disabled citizens on insect walks.

— HOW TO STALK TIGER BEETLES

TIGER BEETLES, MOSTLY diurnal members of the family Carabidae, are charismatic and colorful insects. They behave much like Sanderlings and other shorebirds: they run quickly, stop, run again, and fly short distances if startled. Tiger beetles run so fast that they outrun their eyesight. Once they detect potential prey in the form of another insect, they speed after it. At high velocity, their eyes can no longer focus, and the beetle must pause and reorient to its target.

Tiger beetles mostly inhabit open ground. Look for them on sandy or rocky beaches, dunes, and trails through forests and grasslands. They are most diverse and abundant in early spring, and again in autumn, although some species are active only in summer.

When walking in suitable habitat, watch for tiger beetles flying from beneath your feet. Follow them to their next landing place, and use close-focusing binoculars and/or a camera with zoom capabilities or a lens of appropriate length to locate and observe them. Alternatively, simply stop and watch for the insects running on the ground ahead of you. Where there is one tiger beetle, there are usually many.

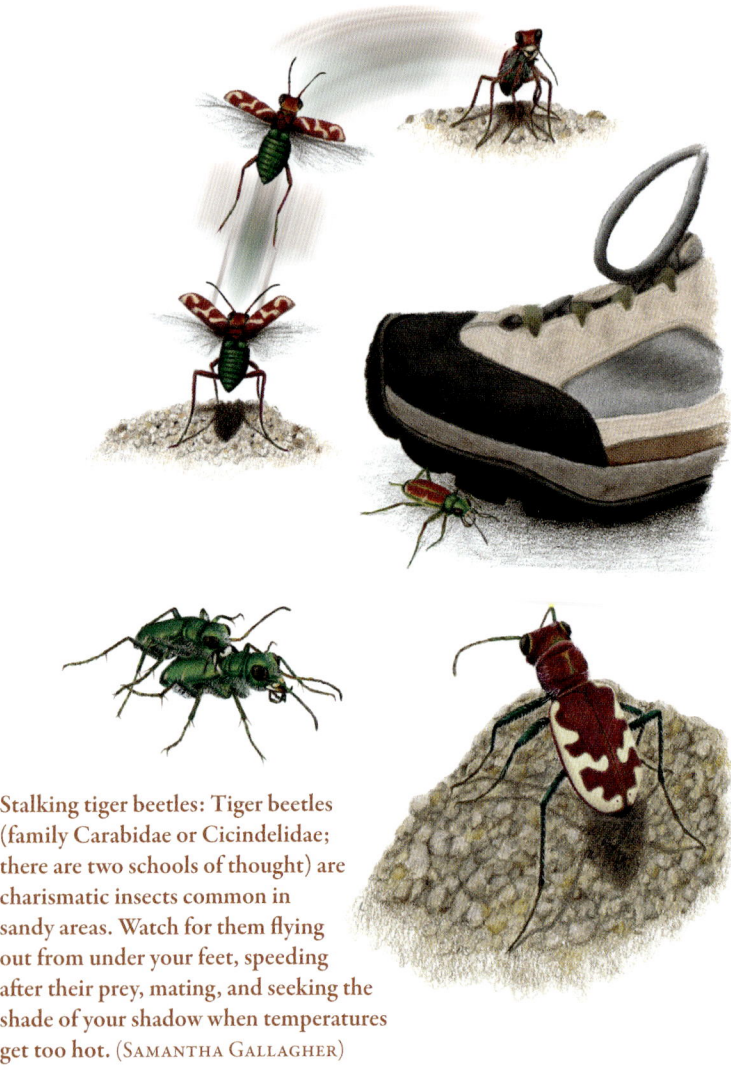

Stalking tiger beetles: Tiger beetles (family Carabidae or Cicindelidae; there are two schools of thought) are charismatic insects common in sandy areas. Watch for them flying out from under your feet, speeding after their prey, mating, and seeking the shade of your shadow when temperatures get too hot. (Samantha Gallagher)

Morning is the best time to get close to tiger beetles, as they are not yet going full speed. As the day heats up, however, they will come running to your shadow, allowing you closer, albeit dimmer, views. They may seek shelter under your shoe.

Most individual tiger beetles you encounter will be alert, standing on tiptoe (to keep their bodies elevated far above the hot sand) or running or flying away from you. Move slowly, and eventually one of the beetles is likely to "hunker down," flattening its body against the ground and relying on camouflage to avoid detection. You can get close-up photos easily under such circumstances.

Tiger beetles are often seen coupled, too, the male atop the female, gripping the narrow part of her thorax with his huge jaws. This is another good chance to get photos, if you move slowly and have patience. Males are reluctant to lose their grip on the opposite sex.

— ETHICS

MOST NATURE RECREATION pursuits are guided by similar codes of conduct to ensure preservation of resources and to maintain healthy personal relationships within the community of participants. This includes staying on designated trails and not venturing onto, or creating, social trails (unauthorized trails created by repeated use of a trail created by one "pioneering" individual, illegally) that compromise the integrity of the landscape and promote erosion. Additionally, bugwatchers are encouraged to follow these guidelines:

— MINIMIZE DISTURBANCE OF other wildlife when pursuing insects. Be mindful of the potential for reptiles and amphibians underfoot, nesting birds in shrubs, and mammal burrows.
— IF YOU TURN over stones, logs, and other objects to look for insects, always return them to their original configuration. Yes, logs may break; simply do your best.
— THINK TWICE BEFORE collecting dead insects. They are still a resource for scavenging creatures and microorganisms.
— BE RESPECTFUL OF other people who may be in the area for different purposes.
— DO NOT TRESPASS on private property unless you have documented permission to do so.
— WHEN IN GROUPS, avoid assumptions about other bugwatchers. Ask them about their level of knowledge and their goals, and ask how you can be helpful to them. Understand that they may not always volunteer their handicaps or fears.
— ADDRESS PERCEIVED ETHICAL violations politely, but assertively. Ask for an explanation of seemingly poor behavior, and voice your personal understanding of rules. Hopefully, that will resolve the issue.
— SEE "DIVERSITY AND INCLUSION," on page 133.

There is increasing attention to ethics around the insects themselves. The Insect Welfare Research Society is mostly concerned with the well-being of insects maintained in laboratories, reared for feed or decorative displays, and reintroduced to the wild. Whether insects feel pain is hotly debated, but the mantra of "first, do no harm" is a good rule for our encounters with insects in the wild, too.

— USING YOUR OTHER SENSES

INSECTS LEND THEMSELVES as subjects of interest to the visually impaired because many species produce sound or emit aromas, and often have intriguing textures. While handling unfamiliar insects is not recommended, a knowledgeable assistant can guide the novice in proper approaches to tactile, olfactory, and auditory examination of specimens.

— LISTENING FOR INSECTS

HUM, BUZZ, WHINE, and *drone* all describe the sound of insects in flight. What would spring, summer, or fall be without the pulse of insect activity made audible? What are nights without cricket and katydid serenades? Close your eyes and tune in to insects with your ears now and then.

The sound of flying insects is frequently a good indicator of an insect's relative size or the quantity of insects present. Low-frequency droning usually indicates a large, slow-moving insect, such as a beetle or true bug. A buzz can mean that a medium-size insect is in your vicinity, maybe a fly, bee, or small wasp. A buzz that stops suddenly usually indicates some type of fly. A whining noise suggests that a small, even tiny insect is close. This could mean a gnat, midge, or mosquito. The wingbeat frequency of such minute species is usually much higher than it is for larger insects. A continuous hum can be characteristic of large numbers of insects, such as a honey bee swarm or hive or a swarm of nonbiting midges, in close proximity.

Aside from these typical descriptions of insect sounds, there are more specific and unusual sounds to listen for. A clattering sound might mean that a dragonfly, or maybe more than one of them, is passing by, especially if you are listening near water. In a field, prairie, desert, or rocky terrain, do you hear a loud crackling sound periodically? That is likely a band-winged grasshopper engaged in crepitation, an in-flight solo act performed mostly by males to get the attention of females (see "Grasshopper Behavior," page xxx).

Walking by a dead, standing tree in the forest, you might hear a gnawing or scraping sound emanating from the trunk. It is possible that a large wood-boring beetle is chewing its way out of that tree. This phenomenon can even occur indoors, as a beetle attempts to liberate itself from a baseboard, beam, or piece of furniture that it was trapped in as a larva. It can take years, even decades, for those beetle larvae to reach adulthood in milled lumber. Tiny deathwatch beetles, family Ptinidae, live in tunnels they bore in wood, and communicate with each other by slamming their faces onto the floor of their tunnel. The resulting ticking sounds, when heard coming from inside homes, were considered omens of impending death in times long past.

A surprising number of insects produce sound by rubbing opposing body parts against one another. This is called *stridulation*. The most familiar examples are crickets and katydids, but other insects use stridulation to attract mates, deter rival males, or in self-defense as a warning sound. Burying beetles in the genus *Nicrophorus* communicate with the opposite sex by rubbing a plectrum (scraper) on the ventral surface of each elytron (wing cover) over a *pars stridens* (file) on the dorsal (top) surface of the fourth and fifth abdominal segments. The result is an audible noise. In crickets and katydids, the scraper and file oppose each other on the "shoulder blade" of each front wing, in the male insects only.

Songs of katydids tend to be raspy in nature, while those of crickets are more musical. The speed of songs varies with the ambient temperature, faster when it is hot, slower when it is cool. The age of the insect and the wear to its stridulating mechanism also influence song speed and quality. Male field crickets and their kin have a loud "calling song" to attract females, a softer "courtship song" to woo her, and a "rivalry song" to fend off competing males. Many katydids and a few crickets have an ultrasonic, or near ultrasonic, element to their songs that may exceed the range of normal human hearing.

Some insects use sound to startle predators into backing off or dropping the insect. Many assassin bugs rasp the tip of their rostrum (beak-like mouthpart) across transverse ridges in a groove on their chest. Some longhorned beetles create a squeaking noise by rocking their heads up and down, raking ridges on their "neck" against the inside surface of their thorax. Velvet ants, which are actually wasps, squeak by means of ridges on the surface of the third abdominal segment rubbing against the inside surface of the second segment. This is an audible warning to predators and, in some instances, part of courtship. Click beetles, family Elateridae, get their name from the loud click produced when they snap a ventral spine into a groove on their chest. This action is so forceful that it can catapult the beetle several inches and/or jar it from the grasp of a predator. Elaterids are bullet-shaped and often covered in a dense coating of fine hairs that makes them slippery to handle.

Those who have hearing loss for high-frequency sounds, which is normal for anyone over the age of fifty, may wish to investigate devices and aids like Hear Birds Again, a phone app developed by Lang Elliott. The open-source iOS software is so far available only for Apple products and is most useful with a binaural headset; it is free to download.

Full throttle: A male Bush Cicada, *Megatibicen dorsatus*, puts his whole body into generating his powerful "song." The special membranes called tymbals are visible at the base of the abdomen, on each side, where it joins the thorax. (ERIC R. EATON)

Most cicadas produce sound through a pair of percussion mechanisms located inside the abdomen of males. Much of the abdomen is hollow, and on each side is a special membrane called a tymbal. The tymbal includes a series of ribs and a plate. Repeated contraction and relaxation of muscles within the tymbal cause the ribs to buckle and relax, creating clicking sounds. These sounds combine, are amplified by the tymbal plate, and resonate within the hollow chamber. The songs of some large cicadas are among the loudest of all sounds in nature. At close range, they are so powerful that it can feel as if your body is being penetrated by a locomotive.

— SNIFFING OUT INSECTS

EMITTING A STRONG odor is another way in which insects defend themselves, and even gentle handling of a specimen can trigger deployment of a scent weapon. Stink bugs, family Pentatomidae, are named for this. Thoracic glands on the underside of the thorax have short channels leading to an evaporative area on the insect's body. Many other true bugs have such aromatic defenses. People have compared the smell to cilantro, or cilantro and ammonia plus skunk. Your personal threshold and appreciation of various aromas will determine whether such scents are good, bad, neutral, or nonexistent.

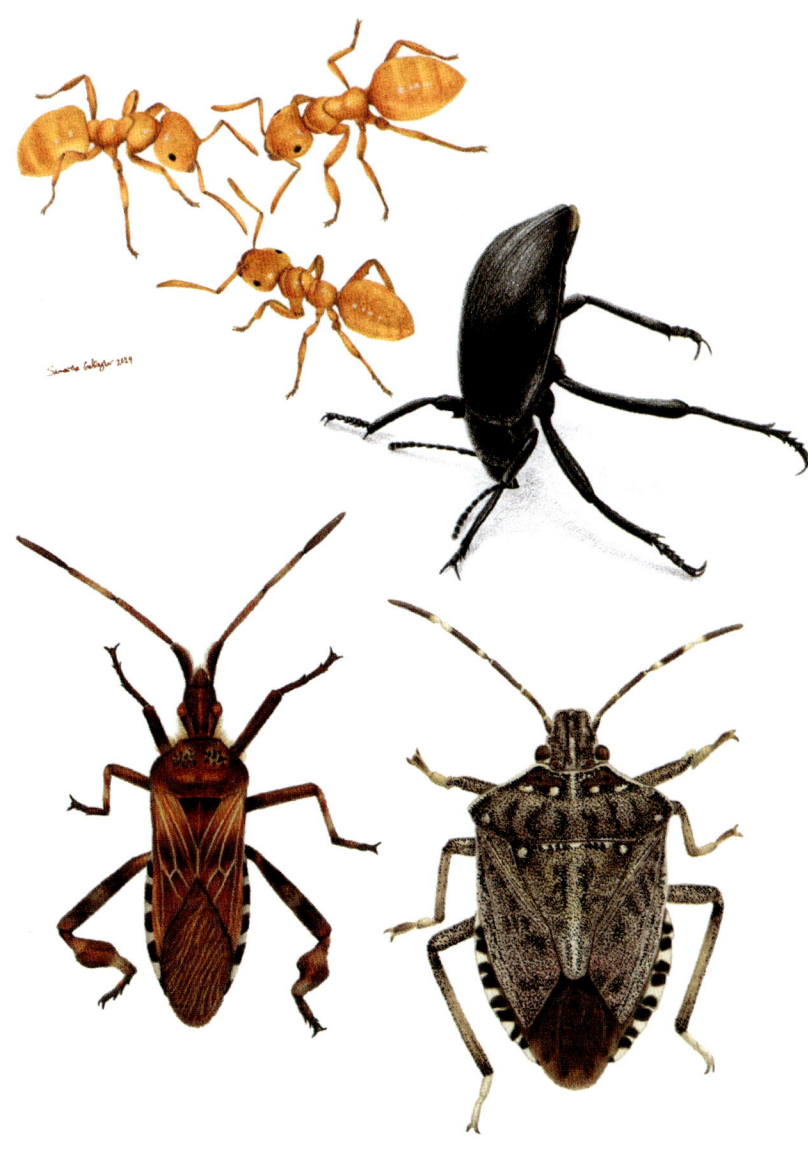

Aromatic insects: Many insects have unique odors, and they are not always pleasant. Clockwise from the top left are citronella ants (*Lasius* sp.), a head-standing beetle (*Eleodes* sp.), Brown Marmorated Stink Bug (*Halyomorpha halys*), and Western Conifer Seed Bug (*Leptoglossus occidentalis*). (SAMANTHA GALLAGHER)

Many ants defend themselves and their nests by spraying formic acid, which has the smell of vinegar. This is true for wood ants (genus *Formica*) and carpenter ants (genus *Camponotus*). The little, black Odorous House Ant, *Tapinoma sessile*, a common indoor pest, when squashed, emits a methyl ketone pheromone, a compound similar to the one that gives blue cheese its distinctive smell. Sweeter yet is the odor of citronella ants in the genus *Lasius*. They smell lemony or have the aroma of a citronella candle. Ants that perish of natural causes emit oleic acid, which has a scent reminiscent of olive oil. This chemical stimulates living ants in the colony to take their dead comrade to the morgue.

Leaf-footed bugs, especially those in the genus *Leptoglossus*, are large, long insects similar to stink bugs, but with flange-like expansions on their hind legs. Their defensive chemicals have been likened to the artificial flavoring or essence of green apple candies.

Many beetles have strong scents. Some darkling beetles, family Tenebrionidae, particularly those in the genus *Eleodes*, are known as stink beetles, head-standing beetles, and pinacate beetles. When threatened, the beetle adopts a near-vertical, head-down posture and emits a foul liquid that quickly evaporates into a stinky gas. Certain ground beetles, family Carabidae, are even more aromatic, and the stench can linger for hours, even after you wash your hands. This is especially true of vivid metallic ground beetles, genus *Chlaenius*, and

Not a tongue: A Black Swallowtail butterfly caterpillar, *Papilio polyxenes*, extrudes a forked gland called an osmeterium, when threatened by a potential predator. The organ emits a repellent compound that discourages further attack. (Samantha Gallagher)

the large caterpillar hunters in the genus *Calosoma*. Hermit flower beetles, large scarabs in the genus *Osmoderma*, have a strong odor of leather when handled.

The alarm pheromones of honey bees are perceptible to at least some people. Caterpillars of many swallowtail butterflies, if irritated, evert from the front of their body a brightly colored, forked gland called an osmeterium. The organ excretes species-specific compounds, frequently similar to butyric acid, which smells like rancid butter.

Certain flies can have an odor. The tiny Mushroom Phorid Fly, *Megaselia halterata*, smells distinctly of anise or licorice.

Wasps usually have a loud wardrobe of black and yellow, orange, red, or white, or perhaps a metallic blue and bright orange ensemble that alerts us to the danger of their potential to sting. Many also give an olfactory warning. The enormous tarantula hawk wasps, genus *Pepsis*, produce a strong aroma when confined in a net. Velvet ants, a type of wasp in which the female is wingless, secrete a pungent odor from glands near their mandibles when they are held. Given the excruciating sting of velvet ants, please take the word of scientists and do not experiment.

— TOUCHING INSECTS

A GOOD RULE of thumb is to not touch an insect unless you know what it is. Some insects are fragile and easily injured, but most are durable, with a dense exoskeleton that is not readily damaged. Larval insects have a more flexible outer armor, sometimes with a dense coating of hairs or waxy filaments. Their health can sometimes be compromised by the oils in our skin, though. Fuzzy caterpillars may look inviting to touch, but some species have venomous spines hidden under those hairs. Adult insects, especially beetles, grasshoppers, and true bugs, are usually less problematic and offer an array of leathery, velvety, smooth, and rough textures that are worth gently exploring. Large, captive species at zoos are ideal for this, under the supervision of the zookeeper or trained docent.

— DO NOT TASTE

TASTING INSECTS IS absolutely off limits under most circumstances. Entomophagy, the consumption of insects by people, is a common practice in many cultures and should not be frowned upon, but doing so safely can be problematic in North America. Many insects are like plants in the ways they defend themselves. They may be full of dangerous toxins, some of which they may derive from plants or animals they feed on. Fireflies, for example, are loaded with lucibufagins, chemical compounds akin to toad toxins. Exceptions to the no-tasting rule are sanctioned public events where specimens are prepared by knowledgeable experts in the science of entomophagy. Even then, there is potential for food allergies or other extreme reactions from your immune or digestive system.

— INDIGENOUS WAYS OF KNOWING

We have to overcome our bias that Western science is superior to
Indigenous ways of thinking.
——————————— LINGUIST DAVID HARRISON, VIN UNIVERSITY, VIETNAM

THE ENTOMOLOGICAL WISDOM of Indigenous peoples and their experience of insects are often ingrained in their eco-centric languages, which are chiefly oral. This lack of a written history or record should in no way be considered less important or less "correct" than the literature of European societies. The overall Indigenous approach to other species is one that views them as relations, as opposed to something apart, to be observed from a distance, without reverence and respect. It is unproductive and disrespectful to further generalize, as Indigenous culture is usually localized, appropriate to a specific society or ecosystem. The heavy hand of colonialism has actively suppressed or obliterated much of this cultural heritage and knowledge. Resurrecting, respecting, and learning ethnoentomology can only enhance the experience of all bugwatchers.

— FIELD GUIDES

INSECT FIELD GUIDES can be a disappointment to anyone familiar with field guides to birds or other vertebrates. Insects are vastly more diverse, and inclusion of all species is impossible. Insects are smaller than other animals, so a mere glimpse is not going to lend itself to identification of the creature. Even with a good look, characters needed for identification may be hidden under wings, behind the head, or at the tip of the abdomen. "Field marks" are seldom as applicable to insects as they are to birds, but the degree of difficulty lessens with practice and experience.

— *"DEAD" GUIDES, LABORATORY GUIDES*

OLDER GUIDES TO insects were designed under the assumption that the user would have a dead specimen at hand. Access to high-magnification optics was often necessary, too. Consequently, they were illustrated with drawings or photos of pinned specimens. The advantage of illustrations is that they often highlight specific physical features that may not be obvious, or even visible, in an image of a living specimen. One drawback is that the insect is usually not in a lifelike posture.

— *"LIVING" GUIDES*

NEWER FIELD GUIDES depict living specimens in postures that one is likely to see in nature. There may also be environmental or behavioral context included

A plethora of field guides: As insects become more popular, field guides to them become more abundant. (Princeton University Press)

in the image, which can be important clues to identifying the insect. The creature may be shown on its host plant or carrying a specific prey species. Ideally, other life stages, like larvae or pupae, are included. It is imperative to note that most insects cannot be identified to species, nor even genus, tribe, or subfamily, from images alone. Please understand that scientific methods of collecting remain crucial to our collective understanding and documentation of insects and other arthropods.

— GUIDES TO SPECIFIC INSECT GROUPS

More popular groups of insects, such as butterflies and dragonflies, have field guides devoted specifically to them. There are also field guides to moths, tiger beetles, fireflies, bees, grasshoppers, crickets, katydids, even caterpillars and flower flies. The sooner you recruit others to the pursuit of your favorite group of insects, the quicker new field guides will be written for that growing audience.

— REGIONAL FIELD GUIDES

Field guides to insects by state, province, or region have the advantage of eliminating similar species that do not occur there, making an identification slightly easier. Such guides may also include specific ecosystem and habitat information that can make it possible to find species on your wish list. Look for such localized guides through state fish and wildlife and parks agencies, as well as nature centers, universities, and museums. Some are produced in PDF formats for free downloads. Bear in mind that climate change is altering species distributions.

— ACCESSORY GUIDES

THERE ARE EVEN guides to insect-related objects such as plant galls, leaf mines, cocoons, nests, and other "signs" that can help identify insects in the absence of the animal itself. These guides are unfortunately few and far between, or available only as PDFs or on websites and other online platforms. This may change if demand increases.

Your best resources for many guides are agencies of the federal government—United States Department of Agriculture (USDA), United States Forest Service, National Park Service, United States Geological Survey—state governments, and university presses, as well as the Cooperative Extension Service. Borrow various guides from your local library to find which ones work best for you. Libraries, nature centers, your local Audubon chapter, and other resources may also have binoculars and other equipment to lend if cost presents a barrier to purchase.

— APPS

APPS ARE DIGITAL software that you can add to your smartphone. These include interactive field guides that may or may not depend on image-recognition algorithms. The quality and effectiveness of apps is variable, and all suffer from limitations that may never be overcome due to the great overlap in physical appearance of even completely unrelated insect species. Seek, a stand-alone app from iNaturalist, has the advantage of anonymity; it does not require an internet connection nor login and collects no personal data. There are no chat or messaging features. Use apps at your own risk, and solicit recommendations from professional entomologists and naturalists. Remember, not every app is compatible with every phone.

— OPTICS

INSECTS ARE SMALL, active, and often difficult to approach. It pays to augment your eyesight with optical equipment that can discern detail at a distance (binoculars) or in your hand (magnifiers). Further, it helps greatly to document your observations and specimens through photographs or videos (cameras, phones).

Make wise choices to be a better bugwatcher. (ERIC R. EATON)

— BINOCULARS

IT MAY SEEM absurd to recommend birding equipment for use in observing insects, but close-focusing binoculars are a sound investment. Close-focusing binoculars allow the user to focus on subjects as close as 8 or 10 feet (2.4–3 meters), sometimes even 3 feet.

When selecting binoculars, pay attention to the two numbers that describe their attributes, such as 7 × 36 or 10 × 42. The first number indicates the magnification value, while the other is a measure of the lens objective (the end of the instrument closest to the subject). There are tradeoffs between lower and higher numbers. The higher the magnification, the larger the image you see, and the better you will be able to resolve details in markings and subtle characters of morphology. Wonderful. Unfortunately, a higher magnification may make it harder to locate the specimen before it once again takes flight (because the field of view is smaller, and therefore you may not have enough surrounding reference points to locate the insect when you put the binoculars up to your eyes). The greater the diameter of the lens objective, the more light is gathered to illuminate the image. In open habitats on sunny days, this is of little consequence, but in the dim light under a forest canopy, or on overcast days, this could make a substantial difference.

Other factors to consider include whether your binoculars are waterproof and how heavy they are to carry. Generally, you get what you pay for, so weigh

Keep your distance: Close-focusing binoculars bring insects closer, in this case literally so. (Joshua Pennington)

the cost against durability, the quality of the image, and the potential need for replacement.

Practice using your binoculars. If you spy an insect, keep your eyes on it, then bring the binoculars up to your face without looking at them. Take your eyes off the bug and you can easily lose track of it.

— MAGNIFIERS

A quintessential tool of entomologists, in the eyes of the average person, is the old-fashioned hand lens, the trademark accessory of Sherlock Holmes. The traditional "disk on a stick" magnifying glass has fallen out of favor, however. Prone to breakage and smudging, and heavy to boot, those awkward magnifiers of yesteryear are now considered antiques.

Today, the preferred compact magnifier (loupe) for field use features one, two, or three lenses that fold inside a protective, oval- or pear-shaped case. Glass lenses are preferable over plastic, but cost more. The most affordable loupes, sometimes known as singlets, are made of a single lens. The curvature of the lens necessary to produce magnification causes distortion of the image

Get closer: The proper technique for using a magnifier loupe, in this case to view the details of the claspers on a male damselfly. (KATHY CARROLL)

seen, especially at the edge of the field of vision. A Coddington lens corrects for distortion by incorporating a diaphragm in the lens. Superior-quality lenses correct for distortion by layering lenses together. A doublet lens has two lenses, and a Hastings triplet has three such layers.

Typically, magnification values range from 3× to 14×. A 10× magnifier is considered standard for nature study in general. The higher the number, the smaller the field of focus. A magnifier with a value of 20× or above is going to have a substantially smaller "sweet spot." Beware of knockoffs, and make your purchase from a reputable scientific equipment supply house. Metal cases tend to be more durable than plastic ones.

To use your magnifier loupe correctly, position it about 1 inch from your eye. In most cases, this is accomplished by simply bringing your thumb up against your cheek. If you are able to manipulate the insect, bring it increasingly close to the magnifier until the image becomes clear. Otherwise, move your head and hand to about 1 inch from the subject. It helps to have good light on the insect.

— CAMERAS AND PHONES

Technology is advancing at such a pace that any generation of products recommended today will likely be eclipsed tomorrow by better models, but here is a primer.

The camera capabilities of many phones exceed the bottom tier of dedicated point-and-shoot and bridge cameras, especially in their abilities to achieve higher resolution and operate in dim light. If you plan to buy a phone to capture images of insects, evaluate the macro (close-up) setting and try it out in the store, if possible. Phones have the advantage of being a device you always carry with you, perfect for when you stumble upon an insect when you least expect to. Phones are also compatible with apps used to share images immediately in social media and/or to identify a mystery critter.

Bridge cameras and "super-zoom" cameras are an excellent choice for the beginner. One can use the zoom feature to focus on an insect several feet away, then switch to the macro setting for a more cooperative specimen at close range. Avoid using digital zoom if you want quality images. Get closer to the subject instead. With bridge cameras, there is no need to switch lenses or dedicate your field time to shooting insects exclusively. Bridge cameras work well for flowers, fungi, birds, reptiles, mammals, and scenic photography, too. There are drawbacks, however. Most do not produce high-resolution images for reproduction in hard copy form because the camera sensor has fewer megapixels than higher-end cameras. Bridge cameras are improving in their ability to perform well in dim light, but they remain inferior to DSLRs and even phones.

DSLR cameras are digital single-lens reflex cameras, used by professional photographers. They are expensive, though most cost is incurred in the purchase of multiple lenses for multiple subjects and circumstances. If you strive for anything beyond documentation through images, DSLR cameras are the way to go, because higher resolution, and the ability to shoot in RAW format, means more pixel density and data for cropping and editing without compromising image quality. Even DSLRs, and lenses for them, are being phased out in favor of mirrorless cameras. (Traditional DSLRs use a mirror to bounce the incoming image onto the viewfinder. When the shutter is pressed, the mirror lifts so that the sensor records the image. Mirrorless cameras permit direct reception of the image from the lens to the electronic viewfinder. What you see in the viewfinder is what you get in the photo. The noise from the flipping of the mirror when you press the shutter is eliminated in mirrorless cameras, allowing them to be quieter.) Focus-stacking capabilities represent another major advancement. (Focus stacking means using the camera to take several images in succession at different focal lengths, such that, in editing later, they can be

THE POCKET CAMERA
MY phone can do WHAT!?

STRENGTHS

Portable & Versatile

Size: bugs are huge in comparison to to small phone lenses. This makes bugs look larger than life in your photos.

Wide Angle: snap the bug and its environment!

There's always a new way to look at the world

Hint: many android phones have a pro-mode built in to the native camera to manually control the focus. 3rd party apps are available for iphones.

GENERAL TIPS

Rule of Thirds: Images that are off center are considered more appealing. Place your focal point (ex. eyes) on the thirds lines

Diffuse your Lighting

Take pictures on cloudy days or shade your bug for soft highlights.

Eye Contact: Humans like eyes! Make sure the eyes are in focus.

SciBugs 2024

(NANCY MIORELLI)

fused to produce one image with a greater focus depth than one gets from a single image. The subject and camera must remain motionless, however.)

Lighting is a profound consideration, so select flash attachments and diffusers as carefully as you choose the right camera. Only DSLR cameras typically give you anything beyond a built-in pop-up flash over which you have limited control. Diffusers, even homemade versions, help soften and disperse light from flashes, be they built-in or attached.

When photographing an insect, take your first shot from farther away, then slowly move progressively closer. The priority is not to spook the insect. To counteract shakiness, brace your hand on an object like a tree trunk or stone, or use a tripod or monopod. Remote shutter releases are ideal. A remote shutter release is a switch on a cable that plugs into the camera. It allows the photographer to trigger the shutter without touching the button on the camera, thereby avoiding camera shake and image-blurring. Moving insects are best shot with burst mode, to guarantee that at least one image will be in focus and/or that the insect will be visible when it is not ducking under leaf litter. You may want to switch to video instead and grab individual frames later. Your most artistic shots will happen when you are at eye level with the insect, or even below eye level.

If images are out of focus or too dark, do not discard them in the field. A great deal of "error" can be corrected in post-processing with various software programs. Make small changes, not great ones, and avoid gross saturation when adjusting contrast and lighting.

From the standpoint of ethics, the well-being of the subject takes priority. Manipulating a specimen to slow it down or pose it may be unavoidable. Take care in doing so. You may also find it helpful to have a white casserole dish or other background material as a "studio" that can be helpful as a contrasting background to get maximum detail or impact.

If you share someone else's photo, always ask their permission, and credit them.

— SOUND RECORDING

It is reasonably easy to record insect sounds with a cell phone, but if you want to devote a professional degree of attention to quality acoustics or record sounds unrelated to singing insects, such as insects feeding or walking, it is necessary to invest in additional gear. Bat detectors, for example, can record the ultrasonic portions of insect songs, as those noises are similar to what bats produce in their echolocation behavior.

Ideally, a directional microphone called a shotgun mic is best because it permits great sound while maintaining a bit of distance from the insect subject

Listening: Lisa Rainsong using recording equipment to document the song of a cricket in Ohio, USA. (Brad Bolton)

you are trying to record. Still, any small-diaphragm condenser microphone is adequate. It may be helpful to also confine the insect in a container that helps block ambient noise, though patience is required while the insect acclimates and accepts its enclosure.

— VISUALIZING INSECT SONGS

Insect songs can be depicted graphically, too, as waveforms and spectrograms. In both cases, time is shown on the horizontal axis, usually in fractions of a second. The vertical axis of a waveform measures relative amplitude. Silence registers as a straight line on a waveform. The louder the sound, the greater deviation in amplitude above and below the silent norm. In spectrograms, the vertical axis measures sound frequency in kilohertz (kHz). On a spectrogram, silence does not register at all; the louder the sound, the darker the marking on the graph, and the higher the note, the higher up the vertical axis it registers. Spectrograms thus provide more information about the

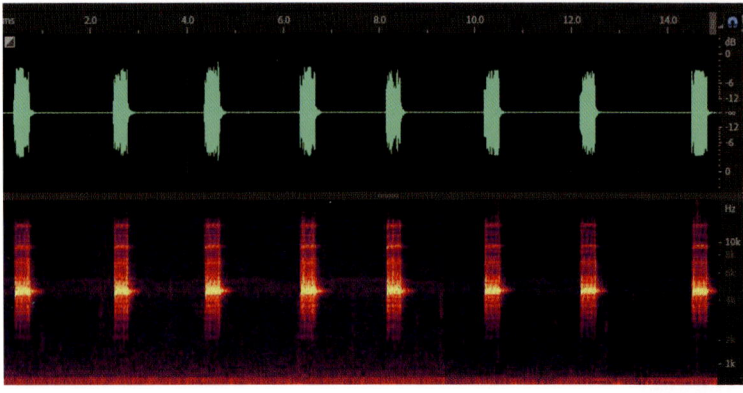

Turning sound into sight: A waveform (top) and spectrogram (bottom) are two visual representations of an insect song. In this case, it is a Jumping Bush Cricket, *Orocharis saltator.* (Lisa Rainsong with Adobe Audition software)

structure and quality of the song. Both waveforms and spectrograms reveal that what may sound like a continuous trill or buzz is, in fact, a rapid punctuation of sound. Hertz (Hz) is a measure of cycles per second; most insect songs register between 2,000 and 15,000 Hz. Human hearing, in our youth at least, ranges from about 30 to 20,000 Hz. Various software programs can produce both spectrograms and waveforms from digital sound recordings.

— NETS

No item of field equipment is as symbolic of entomology as an insect net. When convening with others, nothing says "there are my people" quite like it. A long-handled aerial net comes in handy to secure live specimens for a closer look at anatomical characters needed for identification.

Sweep nets, which are sturdier, short-handled nets, are used to "sweep" grasses, weeds, and light foliage for smaller insects and insects concealed by camouflage or otherwise hidden.

Aquatic nets, often with a triangular or D-shaped frame, are the heaviest of nets, used to sample wet habitats, including the bottoms of lakes and streams. The contents of sweep nets and aquatic nets are typically dumped into a white porcelain tray or pan for easy sorting, though some insects will quickly fly or jump away.

Beating sheets, or beating umbrellas, are durable cloth platforms held beneath a branch while the branch is beaten with a sturdy stick. Strike the branch

Beat it: Using a beating sheet at night to dislodge insects from shrubs and trees can be more rewarding than deploying the technique in daylight hours. (ANTON SOROKIN)

sharply, and it will dislodge hidden insects that will rain down onto the sheet. Again, some insects will quickly run away or fly off, but some will feign death or otherwise remain for you to observe.

You can make your own nets or purchase high-quality, durable nets from biological supply houses. An actual umbrella, held upside down, can suffice for a beating sheet, though insects will hide in the ribs.

— MISCELLANEOUS TOOLS

ADDITIONAL USEFUL ACCESSORIES include a notebook for documenting location, date, time, and other aspects of your observations. A global positioning system (GPS) device, especially one that gives latitude and longitude coordinates, is of great advantage in pinpointing remote locations.

A flashlight is helpful, even during the day. Shine it into burrows, holes, cracks, and crevices to see what insects may be hidden there. Other dark situations, such as cave entrances, areas beneath bridges, the interiors of barns and other outbuildings, and even the forest understory, can benefit from added illumination. Consider a headlamp as an alternative, as it keeps both hands free to manipulate other tools.

Forceps or tweezers are exceedingly useful in extracting insects from crevices, gently grasping venomous species, and manipulating smaller specimens. Soft-touch forceps are less likely to harm insects, especially soft-bodied specimens. A small paintbrush, wetted, is highly useful for gently manipulating

Organism Observed: _____

Corresponding Image/Collection Number(s): _____

Country: _____State/Province:_____

County:_____City/Township: _____

Latitude:_____Longitude: _____

Date: _____Time Start:_____ Time Finish: _____

Name of Observer(s):_____

Habitat Description:_____

Elevation: _____Temperature/Weather: _____

Notes:_____

_____

_____

_____

_____

_____

_____

_____

Sample page from field notebook. (ERIC R. EATON)

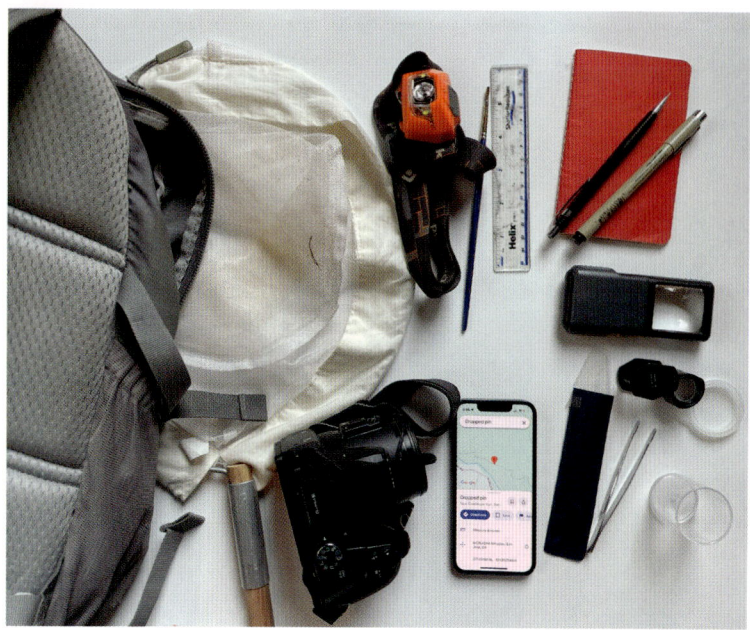

Field kit: Basic tools for observing and manipulating insects include, clockwise from left, a backpack to carry your gear, an aerial insect net, headlamp, small paintbrush, metric ruler, notebook and pens, illuminated magnifier, magnifier loupe, clear plastic vial, forceps with sleeve, mobile phone with GPS app, and digital camera. (AUDREY SAUBLE)

insects and picking up tiny ones. Calipers are the ideal tools for measuring specimens, but at least consider carrying a small ruler. Containers, such as clear plastic vials with snap-on lids, allow for close examination of live specimens. The author often photographs non-climbing specimens through open plastic vials. Placing the white lid under the bottom of the tube helps amplify the flash. Note that many insects can easily climb out of open plastic containers or even glass jars.

— NATURE JOURNALING

A WONDERFUL METHOD for honing your observation skills is nature journaling. One need not be an artist to embark on this journey. The point is to develop an eye for details, and sketching or painting, along with note-taking, is an excellent way to do that. In the words of John Muir Laws, a preeminent

Sample page from a nature journal: Artistic talent is not a prerequisite. Skills in observation and noticing details, relationships between organisms, and habitat context are all enhanced by this kind of exercise. (AUDREY SAUBLE)

contemporary educator, "Nature journaling will enrich your experiences and develop observation, curiosity, gratitude, reverence, memory, and the skills of a naturalist."

There are several books devoted to the art of nature journaling, and many online tutorials. There exist many new and improved, portable supplies, too, like markers and pens, and blank notebooks designed specifically for nature journaling.

— LISA RAINSONG (SHE/HER)

HI! I AM a retired music theory professor, so it may come as no surprise that my interests as a bugwatcher are primarily in crickets and katydids. I did not start seeking them in earnest until I was in my sixties, but in these last ten years, I have learned most of our northeastern Ohio species by ear and recorded them for others to learn and enjoy. Now I present educational programs on singing insects throughout Ohio and have created an online field guide for them.

My equipment includes a shotgun microphone and other recording equipment, and a camera with a macro lens. In doing formal surveys, I pick routes that take me into the likely habitats of the insects I'm most interested in observing. This may change throughout the late summer and fall season, as some insects mature earlier than others. I choose trails that will take me safely to those habitats, as this work is done after dark. I make species lists as I go. Meadows, edge habitats, and wetland edges seem to have the most diversity of singing insects.

Lisa Rainsong
(WENDY PARTRIDGE)

As a woman, personal safety is a big concern. I walk these areas in daylight first, for familiarity. I feel safer when my wife, Wendy, joins me, and she is also a good bugwatcher and listener. I make sure I have multiple flashlights, headlamps, and extra batteries. I use multiple weather apps on my phone, so I can be alerted to incoming severe storms and lightning. I avoid locations that are after-dark hangouts for locals. I have the numbers for the park rangers in every district I survey. If the preserve is not open after dark, but I have permission to be there, I call the local police or park ranger when I arrive and when I leave.

I keep the gas tank full and carry extra food and water in the car. *If anything or anyone ever makes me feel uncomfortable, I leave.*

I get most excited by what I can learn from, and contribute to, the greater community of scientists and enthusiasts. Documenting northerly range expansions of some species due to climate change has been a revelation.

To the beginning bugwatcher: I would simply invite you to slow down to a crawl and start looking at everything you see. Observe not only the beauty of insects, but study where they live and what they do—what they eat, how they mate, what plants they may live on …. then pick a species or a group that particularly enchants you and learn everything you can about them. One insect will lead to another … and another …

LISA'S WEBSITE: https://www.listeninginnature.com/

CHAPTER 3

FINDING INSECTS

DEVELOPING SKILLS FOR detecting insects takes time and is often a matter of trial and error. Adult and immature insects may occupy wildly different habitats and niches, and may be active at different seasons or different times of the day. Evaluating habitats for their potential for insect diversity helps increase the potential for success in seeking a given insect. Camouflage may be the ultimate challenge to finding insects, as they may masquerade as bits of dry vegetation, twigs, leaves, and even bird droppings

— SPOTTING AN INSECT

Once you see one, you'll see them everywhere.

LAURA SCHARE

THE GREATEST DIFFERENCE between watching vertebrates and watching insects is the size of the subject. Train yourself to think small, even tiny. Most insects are built to escape the attention of predators—they are cryptic as well as minuscule. Ones that announce their presence with bright colors are usually

Be the bird: Birds like this Chipping Sparrow are tuned in to the movements of insects like this yellowjacket. Adjust your search image to a smaller scale, and be alert. (ERIC R. EATON)

armed with venom or are toxic in other ways. As with birds and other verte-brates, not every insect can be found in every habitat, during every season, or at any time of day. Do a little research before heading out, and learn where, when, and how to find insects that are of particular interest to you.

To get started bugwatching, simply staring at the landscape is an under-rated exercise. Sitting or standing and gazing near or far has many rewards. Camouflaged organisms will begin to resolve themselves from tree trunks, grass-blades, flower blossoms, and sand or soil. Slight movements will become discernible, and fast-flying insects will capture your attention such that you can follow them.

Birders transitioning to bugwatchers lament that they cannot look down and up at the same time. If you want to see aerial, arboreal, and ground-dwell-ing animals, split the difference. Look at the horizon and let your peripheral vision catch movement both above and below you. Maybe it will be a butter-fly's shadow or a jumping grasshopper, but it is possible to be aware of animals that are both above and below you.

It is best to pick a promising situation or location to begin flexing your pow-ers of insect observation. A patch of flowers is a great place to start; so is a mud puddle in an otherwise dry landscape. A sunlit patch in a forest is another location where insects tend to gather. Gazing at leaves of shrubs where field meets forest will deliver basking butterflies, bees, and wasps, as well as herbiv-orous beetles, caterpillars, and other insects.

Practice noticing movement and following it. Learn the ways insects avoid your vision. Grasshoppers and leafhoppers will quickly shift to the opposite side of a leaf or grass-blade. Beetles will scurry under leaf litter. Slowly reach-ing behind the insect may coax it back into view. Gently lifting a leaf on the ground might reveal the beetle, now sitting motionless.

It is best to assume that any small object is an insect until proven otherwise by close inspection. Each leaf could hide an insect in plain sight or beneath it. Turn boards and stones, but turn them back once you have taken a peek. Turn larger ones toward you to block any potential strike from a venomous snake or stinging scorpion.

— CAMOUFLAGE

Through camouflage, it is possible for even large insects to escape no-tice. That is why slowly moving your gaze over vegetation, up and down tree trunks, and over the ground is a good idea. An insect may even give itself away by moving if it believes you have left the area or that you are a stationary object in the landscape. Insects frequently alight in front of people who are standing still. Sometimes they land *on* people. This is especially true for flies, butterflies, and sweat bees, for various reasons related to the insect's appetite or territorial imperative.

Camouflage includes mimicry of inanimate objects, from pebbles and stones to shards of bark to bird droppings. Stick insects look like twigs or grass stems. The caterpillars of geometer moths, popularly called inchworms, are also twig and tendril imposters, adopting a stiff, motionless posture jutting out from a vine or stem until danger has passed. Some treehoppers are thorn mimics. Katydids look like leaves. Some grasshoppers are so perfectly mottled that they resemble a collection of sand grains. Lichens are a favorite model for mimicking insects, too.

— WHEN TO FIND BUGS: SEASONS

WE THINK OF insects as most abundant in spring, summer, and fall, but of course they are present year-round. They may be passing the winter as eggs, larvae, pupae, or even adults, simply hidden from our view. In some places, there are winter-active adult insects. Mourning Cloaks and other species of anglewing butterflies pass the winter as adults and may be seen flying on warm winter days.

Signs of insects are ever present, but more obvious once foliage has fallen. Galls appear as swellings on twigs, leaves, stems, and other parts of vegetation. Cocoons of moths may adorn bare branches. You never knew the yellowjacket nest was in that tree, but there it is, abandoned now.

Insect species with only one generation each year at a given location are called univoltine. Many insects are bivoltine, with spring and fall generations

Commonly observed insects at different seasons of the year. (ERIC R. EATON)

SEASON				
Early Spring	**Late Spring–Early Summer**	**Mid-Summer**	**Late Summer–Early Fall**	**Late Fall–Winter**
Bees	Everything!	Typically diverse, but lull in abundance	Grasshoppers	Beetles
Flies			Katydids	True Bugs
True Bugs	Moths		Crickets	Lacewings
Tiger Beetles	Butterflies		Mantises	Some Moths
Some Wasps	Caddisflies	Dragonflies	Wasps	Some Flies
Grasshoppers	Some Dragonflies	Katydids	Bees	Barklice
Some Stoneflies		Cicadas	Flies	Gall Wasps
Some Butterflies		Antlions	Tiger Beetles	Some Stoneflies
			Some Dragonflies	Snow Insects

(TYPICAL INSECTS)

of adults, or are multivoltine, with more than two generations per year. This is where a little online or book research can help you determine when to go looking. Miss your window of opportunity and you may have to wait until next year.

— OFF-SEASON ACTIVITIES

Try visiting an insect zoo, butterfly house, nature center, or museum. To avoid crowds, see if you can visit in off-hours or get a behind-the-scenes tour. Most facilities are happy to oblige, if not eager to reward someone for their interest in insects.

Universities have collections of pinned specimens, but often captive live specimens, too. Viewing collections can help you discern details that will improve your ability to identify the living version when you encounter it. University libraries also have a large selection of books about insects, as well as scientific journals that offer more detailed information yet.

You may also need to catch up on curating and posting images from your prior days in the field. Take time to assess what new equipment you will need for the upcoming field season, too.

— WHEN TO FIND BUGS: TIME OF DAY

There are 24 hours in a day, and insects are active during every one of them. Diurnal insects are easiest to find, but crepuscular species are active mostly or exclusively at dawn and/or dusk. Then there are nocturnal insects that require lights and/or good hearing to locate.

The after-dark soundscape is something to behold, especially in eastern deciduous forests and in wetlands, where a great diversity of katydids and crickets can be found. You may be unable to hear some species because the frequency at which they sing exceeds the range of the human ear. Many singing insects are expert ventriloquists that bounce their songs to throw off predators, making them nearly impossible to locate. You can record the insect songs you hear and play them back later; then try to match them with online libraries of audio files of singing insects.

Insects, being cold-blooded, take time to warm up, and going afield in early morning is a great strategy for finding insects that are awakening from slumber before they become too active to approach or follow. Some solitary bees sleep on or inside flowers. Some solitary wasps pass the night in loose clusters or even tight balls of many individuals, mostly males. They start settling down around dusk, so it is sometimes easier to find them when they are actively seeking a roosting spot.

Many desert insects are crepuscular or nocturnal, in order to avoid overheating during the day. If you go looking, be aware that rattlesnakes may also be on the prowl.

— WHERE TO FIND BUGS

INSECTS ARE LITERALLY everywhere. That said, many of the most elusive species are restricted not only by the seasons and time of day, but by habitat. Endangered and fragile habitats such as bogs, fens, caves, dunes, and remnant prairies offer the bugwatcher a special challenge. Most of these habitats are remote and heavily protected by government agencies, nonprofit entities, or private citizens. Permits are often required for accessing them. Their locations may be known by only a handful of people. It is best to get to know those people, prove you can be trusted, and go afield with them. The fewer visits and human disturbances such ecosystems endure, the better.

Two entomologists explore an oak savanna in southern California. Who says bugwatching doesn't have its scenic rewards? (KIM MOORE)

Close to home: Your local park can be a great place to start observing insects. Here, a Leda Ministreak, *Ministrymon leda*, rests by a fountain at Fort Lowell Park in Tucson, Arizona. (ERIC R. EATON)

— URBAN INSECT HUNTING

CITY DWELLERS AND suburbanites can easily find insects close by or within walking or biking distance. Good habitats include arboretums, the grounds of zoological parks, cemeteries, golf courses, vacant lots, and railyards.

Streets lined with tall buildings are (un)natural flight corridors for insects like butterflies and dragonflies. Lights attract moths and other insects during the night, so watch for those still clinging to windows and doors the next morning. Look in building alcoves. Planters on sidewalks and flowerbeds around buildings and parking lots usually feature bright blooms that draw bees, butterflies, and other flower visitors. Water features are usually home to a diversity of aquatic insects. Fountains frequently contain water boatmen, backswimmers, diving beetles, and dragonfly naiads. The waterfront promenade in many cities is an attractive habitat.

Fences and fence posts are used by insects for basking and hunting for prey, and are convenient perches for mating couples. They are also a surface on which to molt. The exteriors of buildings serve insects similarly to fences.

On the fence: Many insects, like this female Carolina Mantis, *Stagmomantis carolina*, take advantage of human structures as places to hunt, mate, rest, and groom. (ERIC R. EATON)

Stored-product pests like meal moths can be found in the pet food aisle at the grocery. The library may have silverfish and booklice but, if so, alert a librarian. Ants, cockroaches, and bed bugs are found everywhere, including movie theaters, public transit, and any other place with serial human occupancy. Termites, deathwatch beetles, and other boring insects are busy consuming our wooden structures, out of sight.

— INDOOR BUG-HUNTING

YOU CAN FIND a surprising number of insects without leaving your home. Living room, kitchen, bathroom, basement, and attic are all potential indoor ecosystems.

Do not be embarrassed by the cobweb in the corner. Compliment yourself for preserving a living pest control agent. See if you can find evidence of insect victims the spider has trapped. Dusty webs, unable to snare prey any longer, can be safely cleaned. Spiders will change "web sites" if they go long periods without success.

Check your pantry. You may need a snack midway through your hunt anyway, but flour, rice, and other grains may harbor unexpected insects. Drugstore Beetles, Cigarette Beetles, Meal Moths, and spider beetles may be feasting on neglected stored products of vegetable origin. Dry, animal-based foods attract the Larder Beetle and carpet beetles, all members of the family Dermestidae. Wool garments in your wardrobe, and wool blankets, furs, and silks are vulnerable to caterpillars of clothes moths and larvae of carpet beetles. Store them in a cedar chest when not using them regularly.

One rewarding source of insect diversity is a light fixture. You may not want to wait for a light bulb to expire before you examine a ceiling fixture or lamp. Insect specimens quickly die in the hot, dry conditions created by a lamp, and become brittle, faded, and gather dust, making them difficult to identify. It's best to check the lights often. Use your magnifier; many specimens will be tiny.

Windowsills are another bug trap, as insects that have blundered indoors will seek a way out by orienting to incoming sunlight.

Houseplants host insects on their foliage and in the potting soil. Overwater and you may attract dark-winged fungus gnats (family Sciaridae), the tiny black flies that always seem to die in a soap dish. Their larvae feed on rotting roots and decaying vegetable matter in general. Look for mealybugs and other scale insects on the stems and leaves of plants.

Back in the kitchen, fruit flies, more appropriately called pomace flies or vinegar flies, genus *Drosophila* (family Drosophilidae), may be coveting overripe bananas on the counter or a bit of spilled wine. Scuttle flies, like *Megaselia scalaris* (family Phoridae), are easily mistaken for pomace flies, but run, stop, and run more often than they fly. They might be breeding in the sink's garbage disposal.

Your bathroom is likely not free of insects, either. Drain Fly, *Clogmia albipunctata* (family Psychodidae), feeds as a larva on organic matter in drain traps and elsewhere. The adults resemble tiny moths and are also known as moth flies. The bathtub might be literally hopping with tiny springtails, innocuous non-insect hexapods in the class Collembola.

In the garage, shed, or attic, don't be surprised to find mud dauber wasp nests, paper wasp nests, and various other evidence of insect interlopers, plus more spiders. Beetles and other insects may emerge from firewood you store indoors.

The basement, perennially dark and damp, is probably home to camel crickets (family Rhaphidophoridae), true crickets (Gryllidae), or both, scavenging on dead insects, molds, fungi, and other sources of sustenance. House spiders, house centipedes, millipedes, and woodlice (terrestrial crustaceans we also call pillbugs, roly-polies, and sowbugs) are other arthropods that may be present.

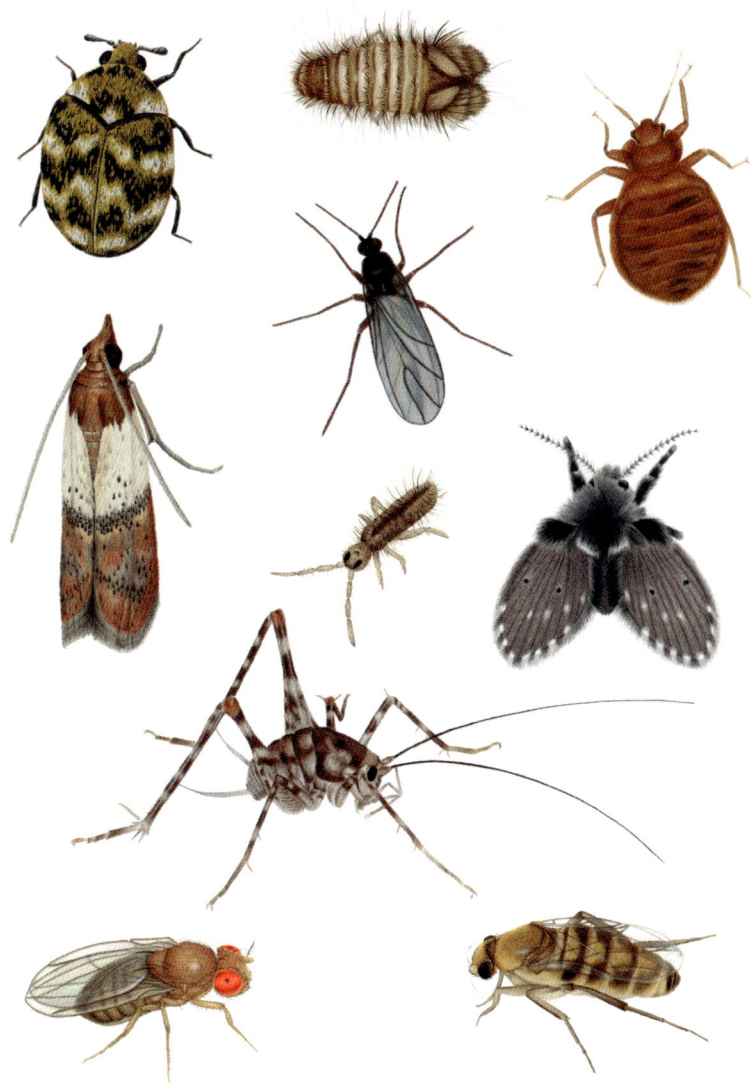

Indoor insects: Among our most common housemates are, clockwise from top left, Varied Carpet Beetle (*Anthrenus verbasci*, adult and larva), Bed Bug (*Cimex lectularius*), Drain Fly (*Clogmia albipunctata*), scuttle fly (*Megaselia scalaris*), pomace fly ("fruit fly," *Drosophila* sp.), Greenhouse Camel Cricket (*Diestrammena asynamora*), and Indian Meal Moth (*Plodia interpunctella*); and, in the center, dark-winged fungus gnat and springtail.
(SAMANTHA GALLAGHER)

In the bedroom, inspect for bed bugs regularly. The same advice applies to the sofa and recliner, bookshelves, bedroom furniture, and baseboards, and to spaces behind wall decor. Adult bed bugs are no larger than the average apple seed. Immature stages are smaller still, and some are nearly transparent. You will likely see signs of bed bugs before encountering the insects themselves. Should you find some, resist the temptation to blame your spouse, roommate, visiting guest, or the tenants of the next-door apartment. Some authorities believe that one out of every four US residences has bed bugs or will have them. *Cimex lectularius* thankfully poses no major medical threats to the average healthy human that modern science is aware of. The biggest problems stem from litigation over infestations and the costs of eradication.

Again, a little digging, in the homework sense, can help you to learn what habitats are best for the insects you want to see. Public gardens may be great for butterflies and bees. The artificial lake or pond at the park is probably good for dragonflies. Glance at the paved bike path now and then for crawling or basking insects or dazed flying bugs that collided with a cyclist.

— TWELVE "MUST SEE" SITUATIONS FOR BUGWATCHING

1. **Public Gardens, Zoos, and Cemeteries**. Botanical gardens and zoological parks typically include landscaping with something in bloom during every season of the year, save winter. Cemeteries, especially older ones, often enjoy similar plantings.

2. **Sunlit Patches in Forests**. Many insects take advantage of illuminated areas in otherwise shady forests for purposes of basking and grooming, and for males to display to females. Watch for hovering flies, bees, wasps, and basking butterflies.

3. **Aphid Colonies**. Aphids, as well as scale insects and treehoppers, exude vast amounts of sugary liquid waste called honeydew. This material attracts a host of other insects. The aphids themselves are also a ready snack for many predatory insects.

4. **Fresh-cut Timber**. Slash piles of freshly cut trees, wind-broken limbs, and recently burned timber attract many beetles that are seldom seen in any other circumstance. Parasitoid wasps and horntail woodwasps also flock to these meccas.

5. **Rest Stops at Night**. Illuminated roadside rest stops, the more isolated the better, are often a treasure trove of nocturnal moths, beetles, and other insects at night. Be careful, and look just beyond the edges of illumination, too.

6. **Mud Puddles**. Water is a scarce commodity in most habitats, especially arid lands, so take advantage of its brief appearance. Stake out mud puddles for visiting butterflies, moths, wasps, bees, and flies imbibing water and dissolved minerals.

7. **Sap Flows**. Wounded trees and plants exuding sap are sure to attract a wide variety of insects, especially if the ooze is fermenting and/or few flowers are in bloom to compete with this carbohydrate source.

8. **Open Summits of Hills, Buttes, and Outcrops**. Male insects, especially butterflies, flies, and some wasps, seek the highest points in the landscape to survey for females carried upslope on the wind. (See "Hilltopping," page 113.)

9. **Cliffs and Cave Entrances**. Many nocturnal insects seek shelter during the day, and diurnal insects want to escape the heat of the midday sun. Cliff overhangs and cave entrances afford protection. Large moths, some dragonflies, and various flies can be among those insects.

10. **Decaying Plant and Animal Matter**. The pile of grass clipping from mowing your yard, the compost heap, or a random deceased animal will all harbor unique insects. This is not an exercise for everyone, and take sanitary precautions when examining such things. Wear gloves, use a stick or other implement to probe, and use hand sanitizer afterward.

11. **Vertebrate Excrement**. Fresh feces from vertebrae animals represents a windfall of nutrients eagerly sought by many insects. This includes the output of livestock as well as wild birds. Break off a twig and use it to see what beetles lurk under fresh manure.

12. **Fungi**. Combine your love of mushroom hunting with entomology and sift through fleshy and woody fungi for various beetles and fly larvae that may be living within.

— WHAT IS A SLIME FLUX?

A SLIME FLUX is a bacterial infection of an injury to a tree or other plant. It typically manifests as a copious amount of oozing liquid that may be bubbling, fizzing, or otherwise showing signs of fermentation. Such gross, weepy wounds are highly attractive to insects, especially some flies and beetles. Most are small, so you have to look closely to see them. Some are found only in situations like the one shown overleaf, such as wounded-tree beetles (family Nosodendridae) and aulacigastrid or sap flies (Aulacigastridae). Pomace flies (Drosophilidae) and sap-feeding beetles (Nitidulidae) flock to slime fluxes, too. Look both day and night.

Slime time: A slime flux will attract many insects that enjoy fermented "beverages." Clockwise from top left are a picture-winged fly (*Pseudotephritis vau*), a Picnic Beetle (*Glischrochilus fasciatus*, with the larva of another sap beetle in front of it), a Winter Ant (*Prenolepis imparis*), two sap flies (*Aulacigaster* sp.), a wounded-tree beetle (*Nosodendron unicolor*), and an odiniid fly (*Traginops irroratus*). (Samantha Gallagher)

— WATCHING WASPS

STINGING INSECTS MAY be the last thing on your list of bugs to observe, but they are fascinating to watch. Most species are solitary and will flee your approach instead of defending their nest. Social wasps are generally placid if you keep a respectful distance.

If you are walking a trail or other area with mostly bare soil and you flush a wasp, then stop. Did the wasp land again a short distance away? Back up a few steps, and see if you disturbed her while she was digging a nest burrow. If so, she will make her way back to the spot and resume her activities. Sit, crouch, or stand (not atop an ant nest!), and watch. Is she a "carrier" like *Ammophila*, that digs with her jaws, then backs out of the burrow carrying a load of soil beneath her head? Maybe she is a "kicker," like *Bembix* sand wasps, using the row of spines on each front leg (tarsal rake) to kick sand beneath and behind her.

You might notice tiny flies flitting and landing near a digging wasp. Those are female satellite flies, family Sarcophagidae, subfamily Miltogramminae, looking for an opportunity to deposit their larval offspring inside their nest. Those maggots will kill the egg of the host wasp and consume the prey the mother wasp furnished for her offspring.

Paper wasps in the genera *Polistes* and *Mischocyttarus* are social. Their nests are uncovered paper combs suspended under eaves and other man-made structures. You can witness the female wasps feeding their larval siblings and each other, cooling the nest by fanning their wings, and other amazing behaviors. If a wasp stands on tiptoe and flares her wings, then it is time to back away until she relaxes again.

— LEAF FLIPPING AND HUNTING CATERPILLARS AT NIGHT

DURING THE HOTTER hours of the day, many insects seek shelter from the heat, hiding beneath various objects, especially on the underside of leaves. Locate those insects by walking beneath low-hanging foliage and looking up; that way, you are not disturbing the insects. Otherwise, try leaf flipping, simply turning over a leaf to see if there is an insect hiding there. Leaf flipping is especially recommended for locating caterpillars, sawfly larvae, many true bugs like lace bugs, and adults and larvae of many beetles. Be careful, as aphids and other true bugs feeding beneath leaves are often guarded by ants that will aggressively attack perceived threats to their "herds."

Caterpillars of many moths feed most actively at night, on the underside of leaves or along the edges of leaves. You can find them with an ordinary flashlight, but many species are well-camouflaged to match foliage and twigs. You

Glowing pillars: Caterpillars can be found rather easily at night because many will fluoresce under blue light. At left are larvae under normal light. At right, they are fluorescing under blue light. (DEBORAH BRUSCO)

may gain an advantage by using an ultraviolet (UV) flashlight. It turns out that chitin, the main structural molecule in the exoskeleton of terrestrial arthropods, fluoresces under ultraviolet light. This is especially true for cuticle (exoskeleton material) that is not heavily pigmented or is thin. Many caterpillars, beetles, ants, and at least one grasshopper and one dragonfly species will fluoresce when excited by wavelengths in the ultraviolet end of the light spectrum; that is, they glow like a neon sign, usually white or green. Do not settle for the least-expensive UV flashlight. Do wear UV-blocking glasses or goggles with this search method. Why do some insects fluoresce? Research is under way, but it may be a way for the creatures to avoid detection by predators, especially birds.

— SNOW INSECTS

SNOW FIELDS IN alpine areas are often littered with slow-moving insects, especially in spring, as insect activity ramps up. The insects become grounded when they fly or are blown onto the cold, accumulated precipitation. It is a great opportunity to find a diversity of species. Even in winter, though, there are insects to be seen that are adapted to the cold.

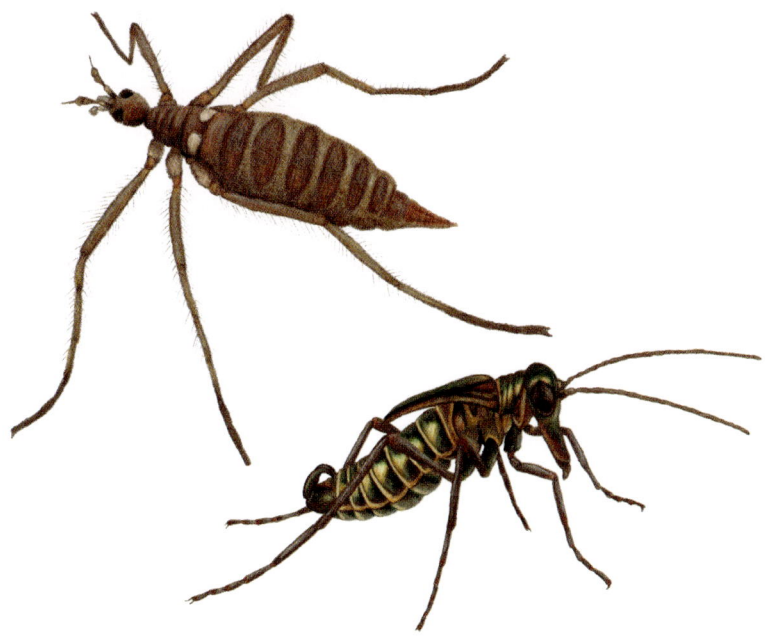

Snow bugs: Some insects, like the female wingless cranefly (*Chionea* sp.) at left and the male snow scorpionfly (*Boreus* sp.) at right, are adapted to thrive in winter. (SAMANTHA GALLAGHER)

Two peculiar snow insects are snow flies and snow scorpionflies. Snow flies are small, wingless crane flies in the genus *Chionea*. They exist mostly in the gap between snow blankets and leaf litter, but are occasionally seen on the snow surface. Equally diminutive, at only 2–4 millimeters, are snow scorpionflies in the family Boreidae. Easily recognized by their long, downward-pointing face, they feed on the leafy parts of mosses and liverworts.

"Snow fleas" are winter-active springtails in the genus *Hypogastrura*. Though tiny, they can form conspicuous masses, especially at the base of trees, where snow tends to start melting first. They graze on decomposing organic matter, mostly from leaf litter. If you see what looks like spilled pepper on the snow, look more closely.

Various small stoneflies in the superfamily Nemouroidea can be found from late fall to early spring, immediately adjacent to clean, fast-flowing streams. Look on rocks, foliage, and bridges for the adults, all of which are less than 15 millimeters in length.

— *CHRIS KLINE (HE/HIM)*

HI! LIKE MANY entomologists and insect enthusiasts, I got my start bugwatching as a small child. My favorite insects to catch were grasshoppers. There's no better drum solo than a grasshopper banging against the plastic lid after it's hopped inside a coffee can.

Today, at over sixty years of age, I am more concerned with documentation than entertainment. I am working, with the help of iNaturalist, on a complete listing of the insect fauna for our 21-acre property in Ohio. My family and I have well over one thousand species so far. We also take our moth lighting rig on the road to different county parks

Chris Kline (CHRIS KLINE)

and festivals, when invited, to teach others the beauty and wonder of moths.

We have landscaped with native vegetation on our property to enhance biodiversity. We do a butterfly transect (sample area) at the beginning of each month, documenting every butterfly we see on the 1-mile trail. We put out bait sticks with our own secret recipe to attract butterflies and other insects. We are fortunate to have friends who excel at identifying moths, so we illuminate light stations at various locations on our land.

The greatest challenges we face are when we venture off our property. Public land in Ohio is scarce, and the parks are crowded, making quality observation of insects nearly impossible. Meanwhile, private property is vigorously defended, even to the degree of gunpoint. Further, as I have aged, arthritis and other health issues have made it difficult to walk long distances like I used to.

As a person of faith, I frequently find myself ostracized by both churchgoing believers and the scientific community. Many religious people consider nature study to be cult-like, worshipping the creation instead of the Creator. At the other extreme, many scientific colleagues are quick to demean those who do not subscribe wholeheartedly to the theory of evolution, even though I have a strong understanding of the principle.

We get the most personal enjoyment out of bugwatching when we find "new to me" species, like the first Xami Hairstreak butterfly we saw. We also get profound satisfaction from opening the eyes of others to the interesting and beautiful insects we share our space with, insects that are frequently overlooked. We do that through our private, for-profit facility, Butterfly Ridge, located on our property.

CHRIS'S WEBSITE: https://www.butterfly-ridge.com/

CHAPTER 4

BRINGING BUGS TO YOU

Employing baits and traps to attract and hold insects, and keeping and rearing insects in captivity, are endeavors that entail pros and cons in terms of labor and ethics. Meanwhile, landscaping with native plants, shrubs, and trees has the potential to rewild your neighborhood and increase biodiversity that supports all forms of desirable wildlife.

— *YOU* ARE THE ATTRACTANT

It does not take much to turn yourself into an insect magnet. Simply standing there and sweating may do the trick. A surprising variety of insects are drawn to the salts and minerals in human perspiration. Aptly named sweat bees, mostly solitary bees in the family Halictidae, frequently

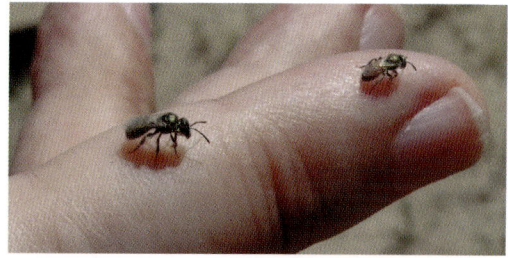

Sweat sippers: Sweat bees, family Halictidae, live up to their name by lapping up your sweat. You can gently brush them off or enjoy watching them. (Eric R. Eaton)

alight on people's exposed arms and legs to sip their sweat. These are small bees, frequently iridescent bronze, copper, or brilliant green. Many types of sweat-seeking flies will make a nuisance of themselves. Those include the House Fly, *Musca domestica*, various tachinid flies (family Tachinidae), flesh flies (Sarcophagidae), flower flies (Syrphidae), lance flies (Lonchaeidae), and others. Even the males of some butterflies, like the Hackberry Emperor, *Asterocampa celtis*, find sweat irresistible.

Some insects are after buried treasure in the form of blood. We all know mosquitoes, family Culicidae, and you should actively repel them, given the disease organisms they can transmit. Other blood-eaters are black flies (Simuliidae), biting midges, also called punkies or no-see-ums (Ceratopogonidae), deer flies, horse flies, and clegs (Tabanidae), the Stable Fly, *Stomoxys calcitrans* (Muscidae), and sand flies (Psychodidae). Tabanids can carry diseases, too, and many people are severely allergic to their bites.

Then there are fleas and lice. While most flea species are specific to other mammals and birds, they will settle for blood from a human in the absence of your pet. Lice are still a scourge, especially head lice. Avoid sharing hats, scarves, and clothing to minimize the risk. Since lice are highly host-specific, there is no possibility of getting them from pets.

What you wear, in terms of clothing and fragrance, can enhance your attractiveness. Bright colors and floral prints can catch the attention of bees and wasps, for example. Various perfumes, colognes, hair-care products, and lotions are even more appealing. There are anecdotes about cosmetic aromas that smell enough like the pheromones of certain moths that the person wearing it will be followed by an entourage of male moths.

Your vehicle can attract insects, too. When parked, its reflective surface is frequently mistaken for a body of water by aquatic insects, and they can crash into it by mistake. When your car or truck is in motion, horse flies and deer flies will surely be in pursuit, believing it to be a large mammal with blood to feed on.

— "SUGARING" FOR MOTHS

IT TURNS OUT that insects like alcohol as much as we do. Entomologists and insect collectors have taken advantage of this for centuries, brewing various concoctions to attract moths in particular. All it takes is a base of stale (flat) beer and a lot of brown sugar. After that, recipes vary greatly by region or even from one cook to the next. If you live in the American Southeast, you might add a fermenting peach to your stew. An overripe banana is a staple of many sugaring formulas. Some folks spike the whole thing with vanilla extract or a favorite fruity liquor, stale orange soda, or honey or maple syrup. Some people

Taking the bait: "Sugaring" for moths is a great way to see some spectacular species. Clockwise from upper left: Ipsilon Dark (*Agrotis ipsilon*), Three-Staffed Underwing (*Catocala amestris*), Short-lined Chocolate (*Argyrostrotis anilis*), Aholibah Underwing (*Catocala aholibah*), and American Bird's-Wing (*Dypterygia rozmani*). (Samantha Gallagher)

swear by rotten watermelon, cut into small chunks and smashed against tree trunks, as the best attractant of all. It is well worth experimenting to see what works best for your particular locale. Do NOT try this method in places frequented by bears.

Combine all ingredients into a jar or pail, and stir until it is relatively thick. The goal is to have a consistency that won't drip or run. Allow the mixture to continue to ferment for at least a day or two, but be mindful that the buildup of resulting gases could explode a vessel with a tight-fitting lid. A paper towel or other permeable fabric, secured with a rubber band, works well.

Find an old paintbrush, a flashlight and/or a headlamp, and some colorful tape, and set off for the set of trees you have selected as your sugaring site. Apply the mixture at dusk, to tree trunks at about eye level, in an area covering about a square foot. By now, the slurry should be plenty aromatic. Avoid trees with heavy ant traffic, and try not to drip at the base of the tree, as ants will send scouts up the trunk. Attach a length of tape or ribbon so that you can find the tree after dark.

Peak moth traffic at the "saloon" is usually between 10 p.m. and midnight, but results may vary. Warm, humid nights with no moon are best, but you will likely have at least a few moths arriving at the bait regardless. When checking, approach cautiously. Given how many moths are drawn to light, you may be surprised to learn that some species avoid it. Keep your flashlight beam aimed below your bait spot, move slowly, and talk softly. Many moths have tymbal organs that are sensitive to sound.

The target species for bugwatchers who sugar with any degree of regularity are the underwing moths in the genus *Catocala*. Many of these are large moths, well camouflaged with broad forewings clad in grays and browns. Their hind wings, concealed when the moth is at rest, are more visually dramatic. Some species have purely black hind wings, sometimes edged in white, but most have underwings marked with alternating bands of black and red, pink, orange, yellow, or white. When discovered on a tree trunk, underwing moths may suddenly expose those bright colors, startling a would-be predator and allowing the moth time to take flight. During the day, you may flush these moths while walking forest trails. They will alight again, some distance away, landing on the far side of another tree trunk.

Once deployed, your batch of bait may be good for another night, but will yield diminishing rewards, if any, thereafter. Watch for beetles, cockroaches, and other insects coming to your bait along with the moths. By day, you may find butterflies, flies, bees, and wasps visiting.

— BAITS

FEEDING THE BIRDS can mean feeding insects, too. As I write this, my partner and I are offering grape jelly, orange halves, and sugar water to migrating orioles and hummingbirds. Instead, there is a feeding frenzy of honey bees, bumble bees, yellowjackets, paper wasps, potter wasps, mason wasps, blow flies, and the odd flower scarab beetle or other insect addicted to sweet liquids. Even birdseed is sometimes infested by moths or granary weevils at some point in the process of production and distribution.

When you mix sugar water for the hummingbird feeder, ideally with distilled H_2O, make an even sweeter batch, about half sugar, half water. Heat to dissolve all the sugar. Spritz a few leaves, ever so gently, to mimic aphid honeydew raining down from the tree canopy. Return later and enjoy the diversity of flies, wasps, and bees at the "bar."

Your uncovered compost heap is an invitation to a wide diversity of flies, beetles, ants, springtails, mites, and other invertebrates that either participate

Not for the birds: A paper wasp (*Polistes metricus*), a potter wasp (*Eumenes fraternus*), a mason wasp (*Monobia quadridens*), and a bumble bee (*Bombus impatiens*) walk into a bar .... Feeding birds sometimes feeds insects, too. (ERIC R. EATON)

in the decomposition process or are in search of tiny prey. Fermentation is a powerful attractant, so decaying fruit, in particular, is wonderful bait. Simply laying a peeled, overripe banana on a stump or other substrate will usually yield something, be it a visiting butterfly or scarab beetle. Remember to look at your baits at night, too.

Because insects navigate their world primarily through the sense of smell, they can be drawn to truly odd, odorous substances. The large, beautiful Banded Alder Borer longhorned beetle, *Rosalia funebris*, found in western North America, may flock to freshly painted buildings. The scent of turpentine and other volatile, plant-based products is frequently employed by entomologists sampling for bark beetles and other wood-boring species.

— TRAPS

Traps are employed mostly by professional entomologists to collect specimens that will be preserved for scientific collections. They are therefore designed mostly to kill insects. A little modification and/or more frequent attention can make them useful to the bugwatcher.

Pitfall traps are miniature versions of the larger hazards one sees in adventure movies where the heroes fall into a camouflaged pit. Simply sink a small

container—like a yogurt cup, margarine tub, or cottage cheese container—into the soil so that the rim is flush with the ground level. Nesting two cups will allow you to retrieve one without caving in the hole. Cover with a board or other object that still allows insects to pass underneath, and you are good to go. You may want to punch a couple of drainage holes in the bottom of the vessel(s) so the trap does not flood if it rains. Pitfall traps can be lethal if you do not inspect them frequently: predatory insects will feed on whatever is trapped in them. Some insects climb well enough that they can get out. In "wild" situations, raccoons, skunks, and other mammals will destroy pitfall traps, especially if you use a bait such as rotting fungi, meat, or dung. Check regularly for the best results.

A couple of simple, nonlethal traps act as artificial hiding places for insects that frequent trees. Try rolling a strip of corrugated cardboard, taping it to hold its shape, and placing it for a few days or weeks in the crotch of a tree or among other vegetation. Remove it, and carefully tear it open to see

OPPOSITE PAGE AND BELOW: Simple pitfall traps can be made by nesting empty food containers and burying them flush with the surface of the ground. Cap with a tile or bark slab. Ground-dwelling insects will fall in. Check regularly to prevent flooding from rain and predation of trapped insects by spiders. (ERIC R. EATON)

Malaise traps are tentlike structures deployed in natural flight corridors for insects. The bugs hit the central panel from both sides and are funneled to a high point, where they are trapped in a container. These traps are best left to professional entomologists. (ERIC R. EATON)

what insects now occupy the inside "tunnels." You can also secure a strip of old carpet or heavy cloth around the trunk of a tree and leave it for a few days. Untie it to find beetles, caterpillars, and other insects that have been hiding there.

Malaise traps are large, tentlike structures. They intercept flying insects and direct them upward into one corner of the trap, to a collection receptacle. A killing agent is usually placed in the jar, to prevent specimens from damaging each other. Malaise traps are usually placed in dry streambeds, canyons, and other natural flight corridors. They are prone to damage from high winds and are vulnerable to vandalism or theft. The traps can be homemade, with much labor, or purchased at a hefty price. Given these drawbacks, it is probably best to leave this trap to the professionals.

While looking for insects, you may come across traps that you did not set yourself. These include familiar pyramid- or tent-shaped traps, open at each end. They are usually baited with a synthetic pheromone directed at one particular insect species. State and federal agencies use these in part to survey for the presence of invasive species. A tall, vertical arrangement of nested funnels, terminating in a collecting receptacle and suspended from a tree branch or attached to a tree trunk is called a Lindgren funnel trap. These traps, designed to mimic bark crevices, are used to sample for invasive bark beetles and other wood-boring insects that hide by day in tight nooks and crannies. Vertical, pyramidal structures of purple panels are aimed at drawing in the adults of the Emerald Ash Borer, another invasive species. The panels are coated with a sticky substance to snare the insects. Avoid disturbing any of these traps, as they are important tools for agriculture and forest monitoring. Report damaged or downed traps to the relevant agency, using the information on the trap.

— BLACK LIGHTING FOR INSECTS

HANGING A BLACK LIGHT (UV bulb) in front of a white sheet or other reflective surface at night is a guaranteed way to see a high diversity of nocturnal insects. Add a bright white light for added attractiveness. The black light will keep the insects in the immediate neighborhood. There is still potential

Bug rave? A black-light setup like this one is a surefire way to attract nocturnal party animals like moths, beetles, true bugs, flies, caddisflies, mayflies, and many other insects. (ANTON SOROKIN)

for carnage, as even the vague warmth of the black light bulb will be enough to kill smaller insects. **Do not stare, or look directly, at a black light**, as UV light can damage your eyes. You can always turn on your porch light and see what flies in. Avoid yellow bulbs. The bluer the light, the better, but a white light is plenty effective. Watch where you step to avoid crushing insects under foot. Since frogs and toads often come to feast at sheets and lights, take care to not harm those amphibians, either. Rarely, skunks, even armadillos, may wander over to snack on "your" bugs. Shake insects off the sheet well before dawn. Otherwise, songbirds will dispatch them in short order.

— PHEROMONE LURES

CARDS AND PAPER strips impregnated with species-specific synthetic pheromones are commercially available for purchase. They are generally used in agriculture and forestry to assess population levels of pest species, but anyone can buy them. They are particularly useful for insects you would seldom see otherwise, such as the day-flying clearwing moths, family Sesiidae. Care must be taken in handling and disposing of the lures to avoid becoming the center of male moth attention. Use them sparingly, anyway, to avoid undue impact on species populations.

— CAPTIVE REARING

RAISING INSECTS FROM an egg, nymph, larva, or pupa stage to adulthood can be a gratifying experience, but with varying outcomes. Be prepared for disappointment, even if you have done everything correctly. Nature is not perfect and is frequently complicated.

Immature insects found feeding on a particular plant will require that plant, maybe even that individual plant, in order to complete their development. You can "sleeve" the twig or limb in a cylinder of cheesecloth or fine mesh, tied at each end to prevent escapes, and monitor it regularly. When all the leaves have been eaten and/or when the insects start to wander, they need to move, or be moved, to more food. Coax them on and off a stick, soft paintbrush, or other object.

A caterpillar found wandering over the ground or on a building may be looking for a place to pupate, having finished feeding on its host plant. If you wish to rear it, place it in a well-ventilated container. Add a layer of soil a few inches thick, as some caterpillars go underground to pupate. Also include a twig or stick, a layer of leaf litter, and maybe a popsicle stick. Cover all of your bases so that whatever substrate the larva needs, it will have it close at hand. Try to identify the caterpillar to learn its habits and preferences. Some, like

Up a sleeve: You can observe naturally located caterpillars and other plant-eating insects by covering the branch they are on with a cylinder of mesh fabric, tied at each end. Watch closely, and let the insects roam again if the branch becomes substantially defoliated. (GREGORY S. PAULSON)

tiger moth caterpillars, will overwinter as larvae and resume feeding the following spring.

Bringing insects indoors can be problematic. They need to be exposed to the same cycle of sunlight and darkness they would experience outdoors. Temperature is less of an issue, though you may want to overwinter insects in an unheated garage or shed, provided there are windows.

Remember that insects are subject to diseases and parasitism, and that the eventual outcome of your rearing efforts may be the premature death of your caterpillar, or a wasp or fly, or several, instead of a butterfly or moth. Do not assume that what emerges from a plant gall you have been keeping is the same organism that formed the gall. Most of the time, it will be a parasitoid (a parasite that invariably kills its host) or inquiline (uninvited guest) instead.

Under no circumstances should you release a pet arthropod at a different location than where it was found. Do take advantage of the expertise at your local nature center, insect zoo, or university lab to help you have a successful experience in rearing insects. There are also many online resources for people who breed insects in captivity. The arthropod pet trade, which is robust and constantly expanding, can be a good place to start. Make sure livestock for sale is ethically sourced.

— BEE BLOCKS

BEE BLOCKS, OR "bee condos," are collections of artificial cavities that solitary bees and wasps will nest in. These take the form of linear cavities that the female insect partitions into individual cells, each one holding a cache of food and a single offspring.

The best designs for such hubs employ wood and stems from trees and shrubs native to your local area. You can drill holes of various diameters into a section of log or a wooden block to a depth of at least 3 or 4 inches. Use sharp drill bits to create a smooth tunnel. Alternatively, bundle hollow twigs from sumac or other twigs with a soft pith (spongy internal tissue). Hang the block or bundle at least 3 feet from the ground and preferably facing south or east. Remember to include a "roof" on the nest to protect it from rain. Some wasps and bees like vertical cavities, so position some hollow twigs that way. Leave erect, hollow stems after deadheading flowers, too.

There is a good deal of controversy and conflicting advice surrounding bee blocks. Some are of the opinion that they are nothing but deathtraps full of fungi, mold, mites, and parasitoids (parasites that invariably kill their host) waiting to kill the bees you intend to make comfortable. Well, so are natural cavities. We do not give female insects enough credit for choosing wisely

Affordable housing: Bee blocks can be a good way to survey what solitary native bees are in your neighborhood, but clean or replace them after each season. (ERIC R. EATON)

where to make their nests. That does not mean we should not properly maintain bee blocks by retiring them after a season or two or by cleaning them using a weak solution of soap and water with a pipe cleaner to destroy any mites or microbes. There is a short window for cleaning, though, between when the last wasp or bee emerges and the first one begins the next season's nesting activities. Again, it is best to consult local experts. Leaving dead, standing trees and other natural sources of dead, solid wood is still preferable, when possible.

Avoid insect "hotels," those conglomerations of bee blocks, hollow twigs, pine cones, dead leaves, and slots for butterflies to overwinter in. They are much more likely to become occupied by social wasps and spiders than the insects they are intended for.

Leave bare patches of soil for ground-nesting bees and wasps, which are the majority of our native pollinators.

— NATIVE PLANT LANDSCAPING

THE ULTIMATE IN beneficial and rewarding undertakings is landscaping with locally native plants. Consider tearing out, suffocating, or poisoning (one time) your lawn and replacing it with a meadow or prairie of native grasses, wildflowers, and herbs. The extent to which you go is up to your own preferences. There are innovative, ongoing experiments in the design of "bee lawns," for example, where clover, creeping thyme, self-heal, and other non-grasses are encouraged. These landscape alternatives typically require less water and fertilizer than conventional lawns.

The most difficult part of such a project may be acquiring an exemption from municipal weed ordinances or a waiver from your homeowners' association. Thankfully, there appears to be a trend toward relaxing strict codes and embracing a wilder landscape, provided it displays some degree of order and sanitation. Your local chapter of the North American Native Plant Society is a great place to start. They often have sales of native plants or can point you to retail outlets where you can buy appropriate species.

Decide what your priorities are in landscaping. Do you want chiefly a pollinator or butterfly garden? Would a water feature be desirable? Maybe include a retaining wall that could serve as a hibernaculum for vertebrate animals, too? How about flagstones, rocks, or log sections for insects to bask on? What about pathways for both maintenance and enjoyment? A place to sit and watch what comes and goes? Whether you are a do-it-yourself landscaper or employ a contractor, take time to thoughtfully consider what your ambitions are. Learn what rules, if any, you must follow, and work accordingly. Be prepared for failure. Not everything will work, and weather and climate change can conspire against you. Be patient and persistent. It may take

Bee friendly: In Minnesota, "bee lawns" are a thing. Any way that you can replace your "lawnscape" with native plant diversity is a win for insects, birds, and other wildlife. (ANNE READEL)

years before that milkweed blooms, longer still before you get Monarch butterfly caterpillars. Thank you in advance for your service to our wild friends and neighbors.

When selecting flowering plants, consider the preferences that insects have. The greatest diversity of pollinating and flower-visiting insects flock to flowers that are white or yellow, maybe pale lavender. Many of those blossoms have nectar guides, patterns in the ultraviolet spectrum that are invisible to us, but easily seen by insects. The favored arrangements of flowers are as a composite (small flowers arranged in dense heads that resemble single flowers: fleabane, coneflowers, sunflowers, asters, and related plants), an umbel (a flat or rounded flower cluster: the carrot and parsley family, Apiaceae), and a raceme (a series of flowers on lateral stalks: mints, wild cherry). Why is this? Individual flowers packed closely together offer one-stop shopping, allowing insects to conserve energy. They need only walk from one flower to another. Secondly, most insects do not have the long proboscis that a butterfly has nor the long tongue that many species of bees have. They require a shallow corolla with nectar reservoirs they can reach with a short tongue (many bees, wasps), short proboscis (flies, most true bugs), or chewing mouthparts (beetles, ants). Lastly, most insects like to be able to detect potential danger while feeding, not dive blindly, headfirst into a tubular flower.

Butterflies and moths will also visit flowers that are blue, violet, orange, or red. Certain sulphur butterflies, in particular, seem to like tubular red flowers. They will brazenly bury their entire heads in blossoms, as will swallowtail butterflies and many skippers. Night-blooming plants will be visited by a variety of moths. Moths will also visit flowers during the day.

Planting for the adult, nectar-seeking insect is only half of the equation when landscaping. The larval or nymph stage needs a host plant to feed and grow on. Research has proven that oak trees host the greatest diversity of caterpillars and other insect larvae. It may not be practical, or even appropriate, to plant an oak, depending on where you live and how much space you have. Plants in the rose family, wild cherry, and native willows also feed an abundance of insects. Herbs in the parsley family are food for Black Swallowtail and Anise Swallowtail caterpillars.

— *KRYSTLE HICKMAN (SHE/HER)*

Krystle Hickman
(Krystle Hickman)

Hi! I have always been interested in small creatures in nature. As a toddler, I would spend time in my yard watching insects. We had rows of rose bushes flowering pink every spring where, in my mind, a billion different insects lived. I remember spending hours looking at ladybugs, spiders, box-elder bugs ... anything that crawled on six or more legs.

I've always had a hands-off approach, leaving things as is and not disturbing. As I got older, my interest in insects combined with my love of photography. It became my goal to share the fascinating stories of these wonderful creatures through photos. It's also a way to humanize them so people will care about them.

If I'm looking for a specific species of bee, I'll research that bees' phenology, meaning where it lives, what plants it likes to visit, at what time of day, and its sleeping habits. I'll use apps like iNaturalist to see if anyone has recently seen the plants that the bees have relationships with. Then I'll visit that area about 30 minutes before sunrise to photograph the habitat, look for those plants, and search for any male bees asleep in the blossoms. The females typically show up later in the day. I'll spend all day observing and taking pictures.

In 2019, I started keeping an Excel sheet to log information from sites I visit. It keeps track of what bees are present, the temperature, rainfall, time of day, and other information. I update it annually to see if there are any patterns or long-term trends.

I'm based in California, so a big challenge is the elements. I've been on freezing-cold mountains and in deserts on 117°F days. I've improved my physical conditioning and endurance as a result. Staying safe is another concern. People have made negative assumptions about me and have called the cops a few times. Sometimes a random person will stand behind me or follow me for an uncomfortable period of time. I have dashcams on my car and always keep my phone and Garmin on my person.

Looking for rare bees involves early mornings. From mid-March through November, I'll wake around 1 a.m. and drive 3 to 4 hours to make it to a spot 30 minutes before sunrise. I must get those landscape shots during golden hour and find those sleeping bees.

What's amazing is making a new contribution to science. I've photographed some of the first living representatives of their species, documented new behaviors and relationships, and expanded the known range of a species. Sometimes I'll spend months looking for a bee. Seeing it for the first time is so exhilarating, especially if it's a species that hasn't been seen in 60 or 70 years.

Krystle's website: https://beesip.com/

CHAPTER 5

IDENTIFYING INSECTS

THERE ARE NUMEROUS obstacles to identifying insects, stemming from their resemblance to noninsects and variability in size and/or color, and due to metamorphosis, sexual dimorphism, mimicry, and unique, often complex anatomical features. How an insect behaves and where you find it can be more helpful than its appearance alone.

— LOWER YOUR EXPECTATIONS

FIELD IDENTIFICATION OF insects is challenging. If you are accustomed to using field guides to identify an organism, expecting to compare images to what you are seeing and assuming that everything will be relatively straightforward, then you are in for a shock when trying to identify an insect. To avoid frustration, set your sights lower. Congratulate yourself for recognizing the creature as an insect or maybe even a specific type of insect, like a beetle or wasp. Build on those successes rather than believing you are a failure because you can't tell if it is a ladybird beetle or a leaf beetle. Staying with the learning process is the only way to get comfortable and advance.

— IS IT AN INSECT?

INSECTS ARE ANIMALS. This is not the common knowledge that it should be. "Bugs" are alien enough that many people assume they are their own category of organism. Within the kingdom Animalia, the class Insecta falls within the phylum Arthropoda. Other arthropods are arachnids (class Arachnida) like spiders, scorpions, mites, ticks, and harvestmen, plus millipedes (class Diplopoda), centipedes (class Chilopoda), and crustaceans such as the familiar crabs, lobsters, shrimp, crayfish, woodlice (roly-poly, sowbug, etc.), and related animals. There are also primitive non-insect hexapods (six-legged invertebrates) like springtails, class Collembola.

All members of the phylum Arthropoda share certain characters. They lack a backbone, like other invertebrates. They have a more or less rigid exoskeleton. They are bilaterally symmetrical, the right and left sides usually mirroring each other in form. The body is divided into segments from front to rear, and the legs have several articulations. In fact, the word *arthropod* translates to "joint-footed."

Insects have three major body sections: head, thorax, and abdomen. This alone separates them from spiders and other arachnids. Intact adult insect specimens have three pairs of legs. Arachnids have four pairs of legs

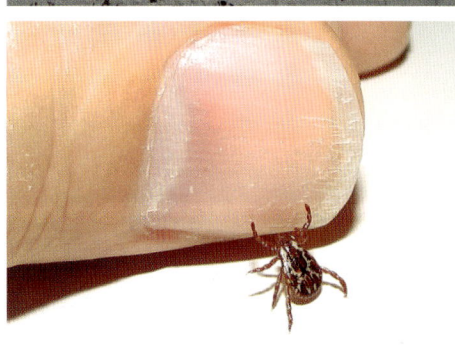

Not insects 1: Many other arthropods resemble insects, but are classified differently.

OPPOSITE PAGE: Arachnids include, top, spiders (like this Black and Yellow Argiope, *Argiope aurantia*), middle, solifuges (*Eremocosta striata* here), bottom left, scorpions (Striped Bark Scorpion, *Centruroides vittatus*), bottom right, tiny pseudoscorpions (unidentified);

THIS PAGE: top, whipscorpions (Giant Vinegaroon, *Mastigoproctus tohono*), middle, harvestmen (*Leiobunum* sp.), and, bottom, mites and ticks (American Dog Tick, *Dermacentor variabilis*).

(excepting larvae of ticks and other mites), and other arthropods have many more appendages than that. Many insects have wings, which no other arthropods possess.

Some insects do *not* resemble insects or any kind of animate creature. Scale insects, related to aphids, lack any discernible features when viewed where

Not insects 2: Top, woodlice (*Armadillidium vulgare*) are terrestrial crustaceans. Middle, centipedes (here a stone centipede, order Lithobiomorpha) are in the class Chilopoda, while bottom, millipedes (barrel millipede, *Cylindroiulus* sp.) are in the class Diplopoda. All these classes, including Insecta, fall under the phylum Arthropoda. (ERIC R. EATON)

they live. They are typically immobile and are covered by a waxy coat that may be hard or soft, depending on the species. They can be easily mistaken for buds or scars, small galls, or lichens on twigs and stems. The presence of wasps and flies can clue you in to the presence of scale insects, as the scales produce liquid waste (honeydew) that those flying insects crave.

— ESTIMATING SIZE

FEW INSECTS YOU see will exceed 3 inches, or 7 centimeters. Entomologists measure insects with the metric system, using millimeters rather than inches or fractions of inches. The overwhelming majority of insect diversity occurs under 10 millimeters. Getting comfortable with measuring in millimeters is helpful when trying to identify insects. Use a ruler, or calipers, when you are able. Bear in mind that, within a given species, there may be great variability in the size of individual specimens. This can be a result of differences related to its sex or determined by the amount of food and nutrients the insect received during its larval stage.

"Big" is relative: This male Northeastern Sawyer beetle, *Monochamus notatus*, is truly large, but most insect species are under 10 millimeters. Having a ruler handy can help in estimating size, though the specimen may not always cooperate. (ERIC R. EATON)

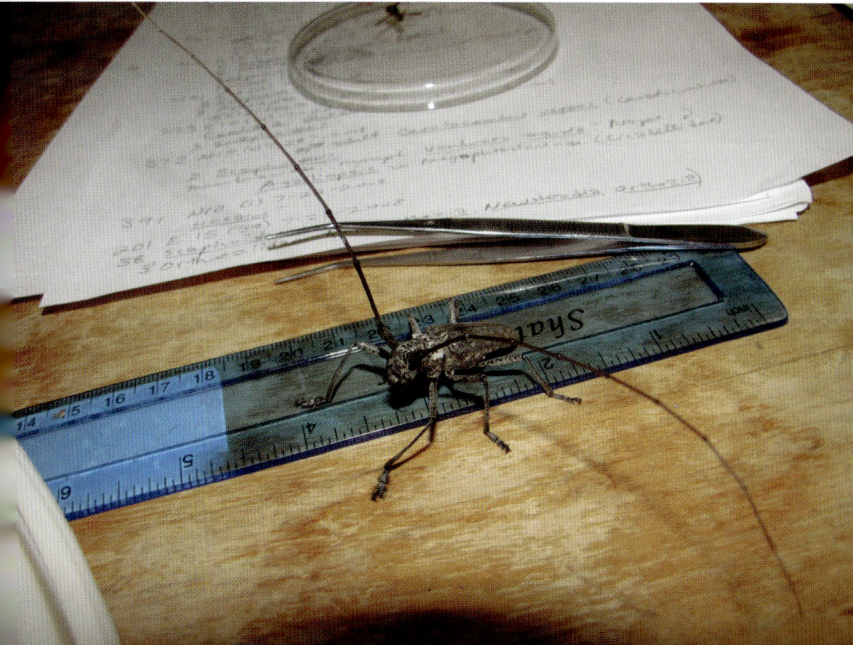

Polymorphic: A "minor" worker Chestnut Carpenter Ant, *Camponotus castaneus*, greets a "major" comrade. The two are members of the same colony, communicating through mutual feeding, known as trophallaxis. (ERIC R. EATON)

— POLYMORPHISM

IN SOME INSECTS—PARTICULARLY social species, such as ants—there exists polymorphism, where individuals may differ markedly in size or in specific physical attributes. Two or more sizes of ants may exist in one colony, for example. "Major" or "soldier" ants are larger individuals that frequently possess larger, more muscular heads to operate larger, stronger mandibles for the tasks of colony defense, prey capture, or related occupations. Smaller "minor" worker ants are dedicated to more delicate tasks, including feeding their major sisters that cannot feed themselves.

Wing polymorphism is another frequently occurring phenomenon. An adult individual may be long-winged (macropterous), while another is short-winged (brachypterous, micropterous) and yet another is wingless (apterous). This is a common circumstance in some true bugs like plant bugs (Miridae) and water striders (Gerridae).

— COLORS

AS WITH SIZE, the colors of insects vary dramatically, even within a single species. An insect species that typically exhibits a balanced pattern of black and yellow, for example, may have some individuals that are mostly black (melanic) or mostly yellow (xanthic). There can be differences in color by habitat and geographic location, too. Insects from arid or warm climates tend to be pale, while insects from cooler latitudes or higher elevations will be darker. Color variation may also be due to sexual dimorphism, age of the specimen, or seasonality. Katydids, green lacewings, and other insects common in autumn will display the same fall colors as the changing foliage.

Many insects, especially butterflies, moths, caddisflies, true flies (order Diptera), and various beetles, are covered in colorful scales that are prone to wear. An older, ragged specimen may barely resemble a fresh one.

Insects that are metallic or iridescent appear to change color depending on the angle of sunlight hitting the creature. That is because the colors are structural, rather than pigmented. Structural colors are the product of bent light, filtering through and reflected by different layers of the insect's exoskeleton. Many insects that are brightly colored when alive fade to dull brown or black in death. In short, the insect you are looking at may not match the one depicted in your field guide, even if they are the same species. One has to look at other, more reliable features of the insect.

— "ALBINO" INSECTS

PEOPLE OFTEN DESCRIBE any white insect they encounter as an albino, but they are usually observing a newly minted insect instead. Immediately after ecdysis (molting), many insects appear pure white, as the new exoskeleton is

Not albino: Most freshly molted insects, like the cockroach on the left, are white, until pigments manifest and the new exoskeleton hardens. True bugs, like the *Leptoglossus* sp. leaf-footed bug on the right, are a striking pink or orange immediately after a molt. (ERIC R. EATON AND KIMBERLY R. FLEMING, RESPECTIVELY)

soft and pigments have not yet manifested. In some other insects, especially true bugs, the freshly molted creature may be pink or orange.

Insects that live their entire lives deep inside caves or in the soil may also lack pigment, but that is not albinism. True albinism is a genetic condition resulting in the complete lack of pigment in species in which the default setting is normal pigmentation.

Many other insects are covered in a waxy white dust (pruinosity) or filaments reminiscent of shredded coconut. These accessories are designed to make them unpalatable to predators, help prevent dehydration, reflect the heat of the sun, or all of the above.

— TENERALS

SOME INSECTS UNDERGO radical color transitions as adult insects. This is particularly true of some dragonflies. The pondhawks, genus *Erythemis*, begin their adult lives as black and green insects that blend seamlessly with the emergent vegetation around the edges of lakes and reservoirs. Females retain this motif throughout their lives, but males become entirely blue as they age, owing mostly to pruinosity.

— BEHAVIOR

HOW AN INSECT moves, for example, can narrow down your list of suspects. If it jumps, it is likely a grasshopper, cricket, katydid, leafhopper or planthopper, or a flea, or a flea beetle. See chapter 6 for a comprehensive treatment of common insect behaviors.

— GEOGRAPHIC RANGE

ONE SHOULD PUT little stock in range maps in books and online resources. The geographic distribution of most insects is poorly known and subject to change. Warming climate is already responsible for northerly range expansions of formerly southern species. Further, species from other parts of the world are constantly being transported accidentally to North America from Europe, Asia, South America, and Australia. That happens in reverse, too. Those records may or may not represent single individuals.

— HABITAT

DID YOU FIND the insect in a meadow? A forest? A desert? Maybe it was in a wetland. Many insects are specific to certain habitats and even certain niches

A dense coniferous forest like this one will support insect species that won't be found in a deciduous forest, prairie, or other habitat. The more habitats you visit, the more insect diversity you will discover. (EMERSON HARMAN)

within a habitat. Some insects are found only on fungi or on specific plants. All of these details are clues that can help identify the insect.

— HOST PLANT OR ANIMAL

THE PLANT ON which you find an insect feeding may offer a vital clue as to the identity of the insect. Many species have a narrow selection of food plants that they prefer. Others are generalists, happy to munch on almost any kind of plant. The same holds true for insects that are parasitoids (parasites that invariably kill their host) or predators of other insects, arthropods, or vertebrates. What part of the plant is it eating? What kind of damage does it do (makes holes, devours a leaf completely)? Make note of the plant if you recognize it. Take images of the plant if you do not know it, so you can identify it later. This is where online communities are invaluable. Someone else will know the plant and identify it for you, and then you have another bit of information that can help identify the insect.

— INSECT ANATOMY

THE LEARNING CURVE for recognizing insect body parts is steep. The same, common feature of an adult insect can look radically different from one order of insects to another. Some segments of the thorax are present in some insects but not others, or they are described with different terminology. This is why most professional entomologists specialize in one particular group of insects—the smaller the group, the better.

The good news is that structural characters are vastly more reliable than color, size, geographic range, and other identification clues. Looking at the basic shape of an insect, the relative size and configuration of the wings, details

Beetle anatomy: Insects like this Fiery Searcher beetle, *Calasoma scrutator*, are a complex collection of segmented parts. At the left is the dorsal (top) side; at the right is the ventral side, or underside. The parts may look different in different types of insects. (SAMANTHA GALLAGHER)

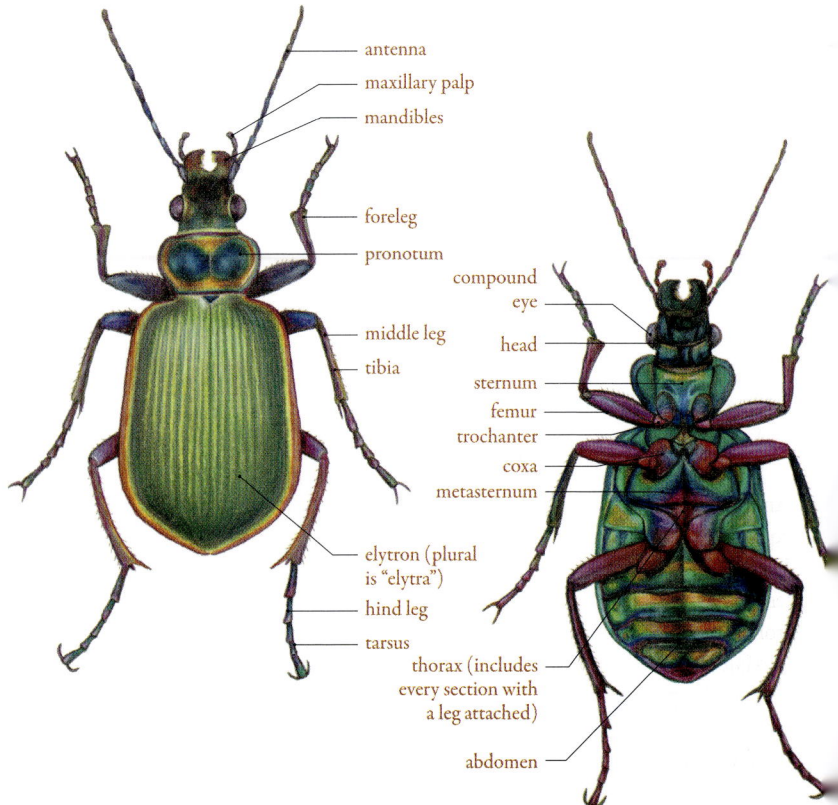

antenna

maxillary palp

mandibles

foreleg

pronotum

compound eye

middle leg

head

tibia

sternum

femur

trochanter

coxa

metasternum

elytron (plural is "elytra")

hind leg

tarsus

thorax (includes every section with a leg attached)

abdomen

of the mouthparts, and modifications of the legs can all help to filter the list of probable suspects.

The three major body sections serve different functions. The head is the centerpiece of the insect's sensory systems, with compound eyes, often a trio of simple eyes called ocelli, plus the antennae, which detect scents and aid in tactile recognition of objects and stimuli. Most insects have chewing mouthparts with opposable jaws (mandibles). True bugs and many flies have piercing-sucking mouthparts that take the form of a rostrum or beak. Weevils have a snout, but there are jaws at the tip. Butterflies and moths have siphoning mouthparts, and most flies have spongelike mouthparts on an extendable "elbow."

The thorax is the locomotion hub, where all legs, and wings, if present, are attached. The interior of the thorax is heavily muscled to operate these appendages. Does your specimen have obvious, membranous wings? If so, you can rule out wingless insects and most beetles, true bugs, grasshoppers, cockroaches, and mantids. Are the hind legs of your insect much larger than the other legs and modified for jumping? Think grasshopper, katydid, cricket, or flea beetle first.

The abdomen contains the digestive tract, excretory systems, and reproductive organs. The abdomen and often the thorax have spiracles, external orifices connected to the internal tracheal network. There may be one or more appendages extending from the tip of the abdomen, which we will identify shortly.

— SEXUAL DIMORPHISM

PHYSICAL DIFFERENCES BETWEEN male and female insects may be so extreme as to lead one to believe they are different species, if not different orders. Popular examples include velvet ants, which are actually wasps, in the family Mutillidae. The females are wingless, usually found running across the ground. Males are fully winged like a proper wasp. Both sexes can be either densely hairy or sparsely haired and more ant-like. Some moths, like tussock moths and the Fall Cankerworm, have wingless females.

In other insects, various appendages are modified, especially in males. Among beetles, the front tarsi ("feet") of the male are often dilated and are equipped with dense pads of setae for gripping the female during copulation. In longhorned beetles, the antennae of males may be more than twice the length of their body; in some genera, like the sawyers (*Monochamus* spp.), the front legs of the male are longer, too.

Male scarab beetles and their kin may be adorned with horns on the head and/or the pronotum (top segment of the thorax). Some stag beetles, like *Sinodendron* spp., may sport a horn, but most members of the Lucanidae family have enlarged mandibles instead. The male stag beetle uses those jaws to battle other males for the right to mate with a female. Male scarabs use their horns for the same purpose.

The shape of the eyes may vary by sex, too. The males of many fly, bee, and wasp species can have eyes that take up their entire face, or nearly so. Those "holoptic" eyes improve the male's ability to spot females and identify competing males that need chasing off.

— SEX ORGANS

LOOKING AT THE rear end of an insect, it is often possible to separate males from females because that is where the genitalia are. Female katydids, crickets, some flies, and various other insects, including wasps, have obvious, spear-like, blade-like, or whip-like organs at the abdomen tip. These are egg-laying organs called ovipositors. Conversely, male insects feature blunt genital capsules at the rear or strange claspers that can resemble grappling hooks or other toothed appendages. Sometimes those claspers are directed upward over the back, as in the scorpion-flies of the order Mecoptera. In other species, like longlegged flies in the family Dolichopodidae, they are tucked beneath the abdomen and directed forward.

Identification of many species of insect can only be accomplished by looking at the genitalia, especially of males. In most cases, this requires dissection of genitalia from a deceased specimen. In the case of dragonflies and damsel-flies, the shape of the male's paired claspers is frequently enough to make a species ID. You may need to net the specimen and examine it up close.

— GYNANDROMORPHS

ON RARE OCCASIONS, a single insect will exhibit both male and female anatomy, including color differences. Such specimens are termed gynandro-morphs. This condition is also seen in other arthropods, particularly crustaceans, and in birds. In the case of insects, gynandromorphy can take two forms, both of them dramatic. Bilateral gynandromorphs are male on one half, female on the other. Alternatively, gynandromorphs can present as "mosaics," with male characters displaying in otherwise female body parts, and vice versa. What causes this chaos? There are several possibilities, the most common being abnormal cell division in embryonic stages of development, resulting in too many or too few sex chromosomes ending up in different cells.

— STING VS. OVIPOSITOR

IT IS UNDERSTANDABLE to assume that any spear-like, whip-like, or blade-like feature projecting from the rear end of an insect's abdomen must be a sting(er) or some other kind of weapon. This is not the case. In fact, if you can see it, ninety percent of the time it is *not* a sting.

Of the many species of wasps, bees, and ants capable of stinging, it is only the females that do so. Fewer still ever use their sting on people. It is social

Not a sting. The spear-like appendage at the rear of this female braconid wasp (*Atanycolus* sp.) is not a sting. It is an ovipositor, which she uses to drill into a log to reach a wood-boring insect larva she will lay an egg on. The wasp larva will feed on the host grub. (ERIC R. EATON)

wasps like yellowjackets, hornets, and paper wasps, ants, and honey bees that deploy their sting in defense of their nests to protect the helpless brood (eggs, larvae, and pupae) within.

A sting is a weaponized ovipositor, a former egg-laying organ that evolved to be connected to a venom gland instead of an oviduct. Eggs are now delivered through a different pathway. Stings, with the occasional exception of some ichneumon wasps, are retracted inside the abdomen when not in use. You do not see a sting until it is firmly embedded and causing you pain.

Males of some solitary wasps are equipped with a pseudo-sting, which may take the form of a menacing hook or trident at the tip of the abdomen. This is part of their external genitalia, not associated with a venom gland. They can stab or prick, but ineffectively.

The longer, spear-like and whip-like "tails" on some solitary parasitoid wasps are true ovipositors. They may also deliver venom, but only to briefly incapacitate the host insect they need to lay an egg or eggs into. They may contort their ovipositors in an attempt to fool their predators into thinking they are weapons, but you have no need to fear them.

Females of many other insects sport ovipositors of other dimensions; they are often knife-like in appearance. Katydids are a prime example; some sport sword-like ovipositors that can match or exceed their body length. These are used to insert eggs into the soil or into plant tissues.

Many smaller insects, such as sawflies and leafhoppers, cut through plant tissues with short, toothed ovipositors in order to lay eggs in plants. Crickets have spear-like ovipositors, as do cicadas and horntail woodwasps. While these appendages can look intimidating, they have a much less violent function: procreation.

— TAILS

WHAT ABOUT LONG, thin, taillike filaments, or shorter, paired appendages on the back of an insect? Mayflies, order Ephemeroptera, typically have two or three "tails," collectively known as caudal filaments. Stoneflies, order Plecoptera, have shorter, paired, antenna-like appendages called cerci. Cockroaches, some termites, crickets, grasshoppers, and webspinners (order Embiidina) also possess cerci. They are mostly used to detect tactile stimuli and movement of air currents.

The shape of the male cercus (singular of cerci) in katydids, spur-throated grasshoppers, and some camel crickets is species-specific and of great help in making identifications. Capturing the specimen for closer examination is usually necessary, though.

— MIMICRY

Mimicry can be the most confusing of all aspects of insect identification. It can make resolution of a given insect nearly impossible, even at the order level of classification. Defenseless flies, beetles, true bugs, moths, and even tropical katydids can successfully mimic stinging wasps or bees. There is also mimicry of a different sort, in the convergent evolution of successful morphology between unrelated taxonomic groups of insects. Mantises, the large, familiar carnivorous insects, and the diminutive mantidflies both evolved raptorial (viselike) front legs for capturing prey, independent of each other. Mantidflies are related to lacewings, while mantises are completely unrelated.

— MIMICRY: FLY VS. BEE VS. WASP

A wide variety of defenseless insects have coevolved to look and behave like stinging wasps and bees, in order to avoid predators. The resemblance can be so strong that even professional entomologists may be duped. How do you tell the difference, even if you get a good photo or a specimen in the net?

The typical template for a wasp or bee is a contrasting wardrobe of black and yellow, white, orange, or red. The usual pattern is one of alternating horizontal bands across the abdomen and bright accents elsewhere on the black body. This is called aposematism or "warning colors," indicative of the insect's ability to fiercely defend itself.

Flies are masters at co-opting this fashion, so we need to look for subtle differences.

— Flies have only one pair of wings, the second pair reduced to knob-like halteres, sometimes obvious in photos, but not always. Bees and wasps have two pairs of wings, but they are joined in flight and function as one pair.

— Flies usually have enormous eyes that dominate their entire head. Some wasps have similar eyes, but most do not; the same applies to bees. Flies may have banded eyes, making them appear smaller.

— Flies generally have short antennae, but they may wave their front legs to imitate the longer antennae of bees and wasps.

— Flies have either sucking mouthparts, appearing as a beak or rostrum, or sponging mouthparts (with specialized structures for sipping up liquids) on a short, fleshy proboscis. Bees and wasps have chewing mouthparts with noticeable mandibles. They can also have a short or long "tongue," a combination of several highly modified mouthparts.

Model and mimic: An Eastern Yellowjacket wasp, *Vespula maculifrons*, top, and a superb syrphid fly mimic, *Spilomyia longicornis*. (Samantha Gallagher)

— Bees are usually robust, rotund even, and densely hairy. Wasps are generally more slender and less hairy. Cuckoo bees, masked bees, and a few other bees can be nearly hairless, though. Scoliid wasps are large, heavy, and bristly, if not hairy. Flies can be any shape, densely hairy or completely naked.
— Hovering behavior is much more indicative of flies than it is of wasps and bees, though some bees, males in particular, are capable of near stationary flight.

Since all three kinds of insects share similar habitats and often visit flowers, you will have ample opportunities to practice making the fine distinctions between all of them. Again, there is no shame in getting it "wrong." After all, deception is the whole point of mimicry.

— MIMICRY OF JUMPING SPIDERS

REMARKABLY, SOME INSECTS mimic the appearance of jumping spiders. Maybe it's not that surprising, considering that jumping spiders are exceptionally keen-eyed hunters that pose a significant threat to insects. What jumping spiders do not usually risk is a head-on confrontation with another jumping spider. Several kinds of flies, especially true fruit flies (family Tephritidae), mimic jumping spiders when viewed from the rear. They have fake eyespots on the rear face of the thorax, and their wings are often marked with bars that resemble the legs of a spider. Some small moths and certain barklice (order Psocodea) have both eyespots and leglike markings on their wings, so they look like jumping spiders in a lateral (profile) view.

Spider spoofers: An Ozark Petrophila moth (*Petrophila hodgesi*), above, sports similar markings, plus eyespots to mimic the arachnid. (ERIC R. EATON)

A true fruit fly (*Rhagoletis* sp.), right, bears wing markings that make it look like a jumping spider, the better to avoid an attack from the rear by an actual jumping spider. (K. LEEKER)

— BIRD POOP MIMICRY

Perhaps the most amusing examples of mimicry are those of insects that masquerade as bird droppings. There are entire categories of moths, for example, that are mottled in tones of black and white, and rest out in the open on leaves during the day. The caterpillars of some swallowtail butterflies pass themselves off as bird feces, too. There are some weevils and other beetles that resemble avian waste. It pays to give that s*** a second look before moving on.

— METAMORPHOSIS

Juvenile vertebrates—most invertebrates, for that matter—are easily recognizable and quickly associated with their adult counterparts. Not true for most insects. Metamorphosis presents a challenge to identifying insects that has been nearly insurmountable for everyone but professional entomologists. Indeed, in many instances, the association between immature forms and the adult insects remains an unsolved mystery. Egg stages, in particular, defy identification, even by veteran entomologists.

The larva, or nymph, stage generally requires several molts. The interval between molts is called an instar. An individual newly emerged from the egg is in its first instar. It is often difficult to determine what instar a given specimen is in, although in nymphs of flying insects that undergo simple metamorphosis, the later instars develop wing pads.

— SIMPLE METAMORPHOSIS

This is the egg to nymph to adult version of insect life cycles. It takes two definitive forms. In primitive insects like silverfish and bristletails, the insect

Simple metamorphosis: The Western Boxelder Bug, *Boisea rubrolineata*, matures in stages that resemble the adult insect, without wings.
(Samantha Gallagher)

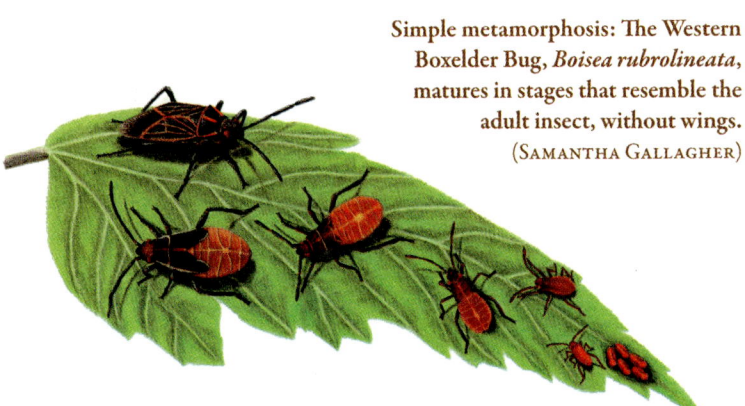

continues to undergo molts after it reaches sexual maturity. These insects are said to be ametabolous. Hemimetabolous insects include cockroaches, termites, mantids, walkingsticks, grasshoppers, katydids, crickets, earwigs, and true bugs, as well as lice, barklice, mayflies, stoneflies, damselflies, and dragonflies. In the case of hemimetabolous insects, the creature does not molt after reaching reproductive adulthood.

— *COMPLETE METAMORPHOSIS*

HOLOMETABOLOUS INSECTS GO through four distinct life stages: egg, larva, pupa, and adult. This is the case for beetles, flies, ants, bees, wasps, lacewings,

Complete metamorphosis: A green lacewing begins life as an egg on a silken stalk, then progresses through several larval instars as a voracious aphidlion. At the end of its larval life, it spins a cocoon in which to pupate, emerging eventually as a lovely winged adult. (SAMANTHA GALLAGHER)

antlions, butterflies, moths, and caddisflies, fleas, and scorpionflies. Complete metamorphosis is considered to be one of the reasons insects are so successful. The adult insects may occupy habitats and niches wildly different from those of their youth. Even their diets may differ radically.

The different stages have different purposes. The larva is the eating and growing stage. It must accumulate enough fat reserves to carry it through the pupa stage, and sometimes the adult stage, too, in the case of insects like silkmoths and bot flies that do not eat as adults. The pupa stage, while usually appearing externally inert, is the reorganization stage, where some youthful attributes disappear and adult morphology manifests. The adult insect is generally the reproductive and dispersal phase, though there are exceptions.

In some moths, like bagworms and tussock moths, the caterpillars disperse by ballooning: issuing threads of silk from glands in their mouths and letting the filaments catch the wind and carry them to greener pastures.

— COMMON NAMES

WHY DO MOST insects lack anything but a binomial scientific name in Latin or latinized Greek? There appears to be a nearly infinite number of insect species, quickly exhausting efforts to assign to each an easily pronounceable English name. There are additional reasons. Historically, common names for insects have been assigned mostly to species exerting an economic impact, especially a negative one. Consequently, many species were given common names that are now seen, and rightly so, as racist or otherwise derogatory. Lastly, there is no standardization of common names. There exists no governing body to approve or reject common names, as there is for scientific names. As a result, "daddy long-legs" can correctly refer to crane flies, harvestmen, or cellar spiders, depending on geographic region.

One helpful convention differentiates true flies of the order Diptera from other insects with "fly" in the name. All true flies have two-word names, like "horse fly" and "bee fly." Insects that are not flies have a singular name: dragonfly, butterfly, firefly, mayfly, stonefly, and so on.

— BUTTERFLY VS. MOTH VS. CADDISFLY

AMAZE YOUR FAMILY, friends, and others by debunking the myth that there is a difference between butterflies and moths. Every rule has at least one exception. Butterflies are beautiful pollinators. Moths are dull, ugly insects that eat wool garments. Rubbish. There are more essential pollinators among the moths than there are butterflies. Only a handful of moth caterpillars are truly

Posing: A moth may sit at rest with wings folded or outstretched, as demonstrated by this Clover Hayworm, *Hypsopygia costalis*. You cannot always rely on your field guide for matching images. (ERIC R. EATON)

pests, and butterflies like the Small White (aka Cabbage White), *Pieris rapae*, are no friend to the vegetable grower.

Butterflies are diurnal, and moths fly at night. No, many moths are also diurnal, and a few butterflies will be active at night. Butterflies are brightly colored, but so are many moths, especially those that mimic wasps or bees and that are advertising their distasteful, or even toxic, nature.

Moths have feathery antennae. The *males* of many moths have those plumose antennae, but by no means does that apply to all moths or even all males. Butterflies have clubbed antennae, but one group of tropical moths does, too.

Butterflies have slender bodies, while moths have bulky bodies. In proportion to wing area, this is frequently true, but again not always. Skippers, a subset of butterflies, typically have heavier bodies and smaller wings, for example.

Moths rest with their wings open, while butterflies at rest have their wings closed. Butterflies often bask with their wings open or send a visual message to a potential mate or rival by opening their wings. Meanwhile, moths can also change the configuration of their wings at will. Many species roll their wings around their bodies, especially in cold or windy weather. The shape of moth wings is usually narrower than that of butterflies, but not always.

The only solid distinction between butterflies and moths is in how the forewings and hindwings are connected or not. Moths have a hook-like bristle, or bristles, called a frenulum, found on the front margin of the hind wing. It connects to a corresponding structure, the retinaculum, located on the hind margin of the forewing. This pairing allows the two wings to function as one in flight. Butterflies do not possess this character. Oops, the silkmoths of the superfamily Bombycoidea, lack the coupling mechanism, too. Since butterflies evolved from moths, it may be more correct to call all the members of the order Lepidoptera "moths."

Caddisflies are easily mistaken for moths and are also attracted to lights at night, sometimes in greater numbers than moths. As larvae, they are aquatic, but as adults, how can you tell them from moths? Caddisflies have their wings covered in hairs, rather than scales, and most species have few, if any, contrasting markings.

— BUG VS. BEETLE VS. COCKROACH

How do you tell a true bug from a beetle or a cockroach? To the untrained eye, they are similar, if not nearly identical. Here are a few key characters to look for:

Antennae. True bugs mostly have antennae composed of a few long segments, though in some cases, they are reduced to hairlike bristles. Beetles have a diverse array of antenna forms, but never filiform (hairlike or whiplike, with no obvious segmentation). Cockroaches have those long, filiform antennae in most cases.

Mouthparts. True bugs have piercing-sucking mouthparts in the form of a rostrum or "beak." Both beetles and cockroaches possess chewing mouthparts with opposable mandibles.

Wings. Adult true bugs usually have wings, though some do not. In the case of winged specimens, the front wings overlap at rest, creating an X pattern over the back thanks to a large triangle near the center of the insect, called a scutellum. Beetles usually have a reduced scutellum, and the hardened front wings (elytra) meet along the midline instead of overlapping. Winged cockroaches have the leathery front wings overlapping, but no scutellum to create an X pattern.

Cerci. Only cockroaches have obvious, paired, taillike appendages called cerci. These usually protrude beyond the margins of the overlapping wings.

opposite: Easily confused: True bugs, beetles, and cockroaches look similar at first glance, but subtle differences in antennae, mouthparts (true bugs have a rostrum instead of jaws), wings, and rear appendages (or lack thereof) can help separate them. Roaches have spiny legs, too. (Eric R. Eaton)

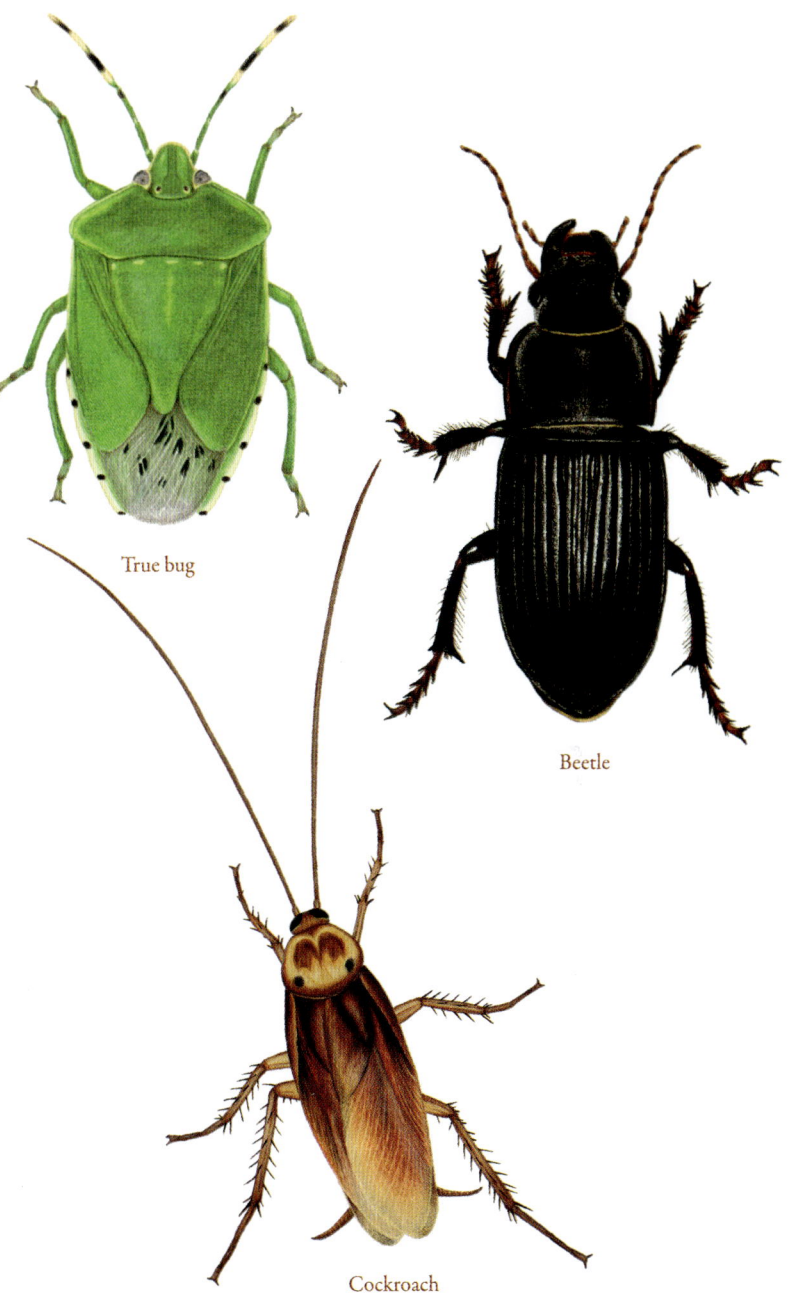

True bug

Beetle

Cockroach

— WHAT IS AND IS NOT A JAPANESE BEETLE

ONE TRAGIC CONSEQUENCE of insisting that nonnative species are pests that must be dispatched at every encounter is that most people cannot properly identify the villain in question. They end up killing a harmless or beneficial organism instead. The Japanese Beetle, *Papillia japonica*, is a case in point. Several other beetles are confused for them: green June beetles in the genus *Cotinis*, the Emerald Flower Scarab, *Euphoria fulgida*, and the Dogbane Leaf Beetle, *Chrysochus auratus*. Even non-beetles, like green stink bugs, both adults and nymphs, can be mistaken for the Japanese Beetle. One must note several specific details to distinguish among insects, not just color and shape.

— GRASSHOPPER OR LOCUST?

LOCUSTS ARE NOT species, but instead represent a physical and behavioral variation of some grasshopper species. Certain grasshoppers, when overcrowded in the flightless nymph stage, will quickly exhaust the local food supply and march to greener pastures. Their sheer density means they literally rub knees with each other. The tactile stimulation triggers a change in metamorphosis from the usually solitary condition into a "gregarious" phase. This results in a different breed of adult than normal: a "locust." With more streamlined bodies and longer wings, locusts are able to disperse over greater distances in search of still more food and more hospitable breeding locations than solitary grasshoppers. They maintain their collective hordes with devastating consequences. The phenomenon of true locusts is confined to the Middle East, southern Europe, northern Africa, and parts of Central America.

Our European immigrant forefathers called periodical cicadas "locusts," because they had only seen true locusts in the vast numbers that periodical cicadas present. Periodical cicadas are unrelated to grasshoppers.

— IS IT A "KISSING BUG?!"

ANOTHER EXAMPLE OF fear-stoked misidentification happens with the blood-sucking conenose bugs, or "kissing bugs," assassin bugs in the genus *Triatoma*. They feed on the blood of mammals, mostly rodents, and are nocturnal in habit. Because some species in this genus are vectors of Chagas disease, caused by a parasite called a trypanosome, there is understandable anxiety associated with any insect that looks remotely similar. Most often mistaken for kissing bugs are other assassin bugs and leaf-footed bugs in the family

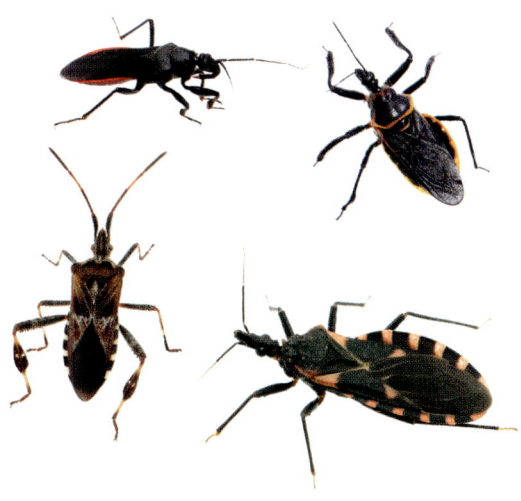

Which is the real Japanese Beetle? Clockwise from bottom right, Japanese Beetle (*Popillia japonica*), Emerald Flower Scarab (*Euphoria fulgida*) in flight, Dogbane Leaf Beetle (*Chrysochus auratus*), and Green June Beetle (*Cotinis nitida*). (John C. Abbott)

Kissing confusion: A real "kissing bug," a blood-sucking conenose (*Triatoma sanguisuga*) in the assassin bug family, is at bottom right. Above it is a bee assassin (*Apiomerus* sp.), with a Black Corsair assassin bug (*Melanolestes picipes*) at the top. At left is a completely harmless leaf-footed bug. (John C. Abbott)

Coreidae. *Triatoma* get the name "kissing bug" because when they feed on sleeping people, they go for exposed parts, like the lips, that are not covered by sheets and blankets.

— METAMORPHOSIS MYSTERIES

EXPLORING PLACES LIKE the edges of ponds and lakes, rocky stream banks, and even the exterior of buildings can produce ghostly evidence of insects past in the form of exuviae. *Exuviae* is plural, *exuvia* singular; the terms refer to the discarded exoskeleton left behind after an insect has molted.

Exuviae can be confounding because they often represent stages in the insect life cycle that are hidden from view. Aquatic insects like dragonflies, stoneflies, and mayflies live most of their lives underwater. Cicadas spend years living a subterranean existence underground, feeding on sap from plant roots.

Ghostly remains: The shed exoskeleton of a darner dragonfly persists on the cattail that the naiad climbed out on. The adult insect emerged and left this behind. The white "strings" are portions of the tracheal system, invaginations (infoldings) of the external cuticle, and are shed with it. (ERIC R. EATON)

It is only when these creatures surface, to metamorphose into adult, winged insects, that we see the shells of their former selves.

Molting most often takes place during the night or at dawn or dusk, so we rarely witness the event. Instead, we are confronted with puzzling objects the following day.

Mayflies are especially interesting because they molt after they reach adulthood. The *nymphs*, or *naiads*, which live underwater, climb a handy vertical surface and molt to a short-lived adult stage called a *subimago*. Fly fisherman call this stage a *dun*; it molts a second time into a full-fledged adult or *imago*. Anglers know these as *spinners*, though that term applies mostly to exhausted mayflies at the end of their lives that are simply sitting ducks on the surface of the water.

Look closely at the exuviae of larger insects and you will see white threads spewing from within. These are part of the tracheal system of the insect, invaginations (infoldings) of the exterior cuticle that carry oxygen into the inside of the insect before branching into ever-smaller tracheoles that are not reinforced with chitin.

Synchronous molts occur in many insects, particularly true bugs with warning colors that spend their youth in loose groups. Since color patterns change with each molt, it pays to be on the same page with your cohort.

There are some insects for which you will never see exuviae because molting occurs inside a nest, a tree, or some other fortress, or because the insect consumes the shed cuticle after it emerges from it. Insects seem to understand that birds and other predators notice exuviae and will search in earnest for the insect that emerged from it. Thus, it pays to eat the evidence.

— OOTHECA AND EGG MASSES

IDENTIFYING THE EGGS of insects is much more difficult than determining the identity of adult specimens. Still, some egg masses are so distinctive that identifying them is relatively easy. You can always place insect eggs in a container and see what emerges.

The eggs deposited by mantises are among the most conspicuous and easy to identify. The eggs are usually laid in neat, dense rows, adhered to each other with a liquid material that hardens into a protective covering that resists desiccation. The resulting mass is called an ootheca. These objects are still sometimes confused with slime molds, fungi, or even oozing foam insulation when they are discovered on a man-made structure.

Eggs of the Wheel Bug, *Arilus cristatus*, a large species of assassin bug, are laid in a distinctive, hexagonal mass, usually on the bark of trees. The emerging nymphs are bright red and black with yellow-tipped antennae.

What is hatching? It helps to identify insect eggs if you witness the babies emerging. First-instar Wheel Bug nymphs are beginning to erupt from the typical hexagonal egg mass at left. (KIMBERLY R. FLEMING) Below left, leaf-footed bugs, *Leptoglossus* sp., are leaving the line of eggs their mother laid. (K. LEEKER)

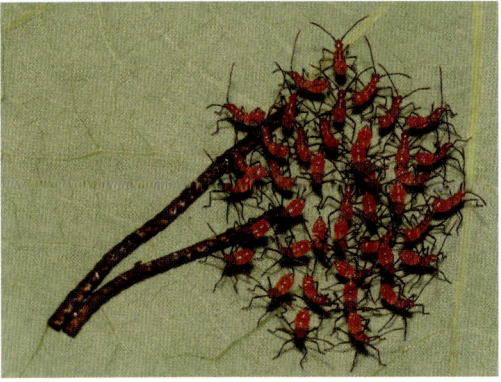

Stink bugs lay small groups of eggs, each often adorned with a crown of tiny spines. The eggs are more easily seen immediately after the nymphs emerge. The young bugs encircle the empty eggs for a short time before dispersing.

Eggs of dobsonflies, genus *Corydalus*, are laid in flat masses covered in a white, chalky coating and are guarded by the enormous female insect. Look for the masses and their remains long after the larvae (hellgrammites) have emerged, on the exterior of buildings near flowing water.

Butterfly eggs are usually egg-shaped and deposited singly on the foliage of the host plant. Look for butterflies that land briefly on leaves, flit to another leaf, and so on. These are likely females laying eggs. By contrast, many moth species lay eggs in large batches. Tent caterpillar moths and some other moths lay their eggs in a ring or band around a twig or stem. Wingless female tussock moths lay their eggs on the outside of the cocoon they emerged from.

If you decide to try to rear insect eggs, remember that there are tiny parasitoid wasps that can develop as larvae inside the eggs of their host insects. You may therefore get the expected insect or, instead, a diminutive, non-stinging wasp.

— COCOON VS. PUPA

HOLOMETABOLOUS INSECTS, THOSE that undergo complete metamorphosis, pass through a pupa stage. This stage usually resembles a near-lifeless blob from which the adult insect will eventually emerge.

The larvae of some insects spin a silken envelope around themselves prior to pupating. This self-generated package is what we call a cocoon. It is not the same thing as a pupa, any more than a sleeping bag equals the slumbering human inside. Moth caterpillars are the most well-known example of cocoon-spinners, but not all moth species bother. Many moth pupae are naked, hidden underground in leaf litter, under bark on trees, and in other nooks and sheltered situations. Meanwhile, sawflies, wasps, bees, ants, lacewings and their kin, and some beetles also spin cocoons.

The term *chrysalis* applies to the pupa of a butterfly, but not any other insect.

— GALLS

A GALL IS a plant growth initiated by another organism. This may be an insect, a mite, a fungus, a bacterium, or something else. Among insects, gall wasps and certain flies, sawflies, aphids, and psyllids are typical gall formers. Gall wasps, family Cynipidae, are tied almost exclusively to oaks and some plants in the rose family. If a gall has a hole or holes, then the occupants have already emerged. Rearing galls can be rewarding because what comes out is often not the insect that created the gall. Many inquilines (uninvited guests) and parasitoids (parasites that invariably kill their host) attack gall makers.

— HOLES BORED BY INSECTS

WOOD IS OFTEN damaged by all manner of boring insects, and the holes they make can give a clue as to their identity. Moreover, many such tunnels are re-used by other, completely unrelated insects. A dead tree offers homes to a high diversity of insects.

The holes you see in a weathered log or a dead, standing tree are usually the exit passages of an insect that was living its larval life tunneling through the wood, often with the aid of symbiotic fungi that digest the cellulose. The shape and size of a given hole can be a clue to the identity of the insect that made it.

Horizontally oval holes are typical exit holes of flathead borers, also known as jewel beetles or metallic woodborers, beetles in the family Buprestidae. Perfectly round, usually moderately large holes are made mostly by roundheaded borers, better known as longhorned beetles in their adult stage (family Cerambycidae). Horntail wood-wasps, family Siricidae, also leave round exit holes.

Small or tiny round holes are usually the work of deathwatch or powder-post beetles, in the spider beetle family Ptinidae. It can look like buckshot was fired at the tree.

All of these holes can be taken over by different kinds of solitary wasps and bees that nest in preexisting cavities. On some occasions, real estate is in short enough supply that one species will usurp the active nest of another species (or even another member of its own species), casting out the egg, larval, and pupal occupants to make way for its own nest. The ready-made tunnels are partitioned into individual "apartments," each one containing a single offspring, along with a loaf of pollen or paralyzed insects or spiders as food, depending on the species of bee or wasp.

— BACKSWIMMER VS. WATER BOATMAN

AQUATIC TRUE BUGS are common in nearly all stagnant bodies of water. Chief among them are predatory backswimmers and the mostly smaller, omnivorous water boatmen. The two are frequently and easily confused.

Backswimmers, family Notonectidae, are triangular in cross-section, flat on the underside (the ventral side). They swim upside down, with their back forming a streamlined keel that enhances their mobility. Long hind legs, fringed with hairs, work like oars to propel them swiftly. The other two pairs of legs are not modified, but the front pair is used to seize prey. A typical rostrum with piercing-sucking mouthparts inflicts a paralyzing bite and begins the digestive process. Backswimmers are mostly open-water hunters, easily observed as they row through the water.

Water boatmen, family Corixidae, are flat-bodied. They swim right side up and spend most of their time on the bottom of shallow bodies of water. Water boatmen usually have a finely reticulated pattern of black and brown, making them difficult to see. Their hind legs are modified similarly to backswimmers. The middle legs help anchor them to a substrate. The tarsi of their front legs are spoon-shaped, the better to shovel food particles to their mouths. The rostrum is greatly reduced and inflexible. These insects also possess a unique grinding organ that allows them to ingest solid food. Their diets are complex and not fully understood. They cannot bite people, unlike backswimmers.

— CATERPILLAR OR SAWFLY LARVA?

THE DEFAULT INTERPRETATION of a wormlike insect feeding on foliage is that it must be a caterpillar. Not so fast. The larvae of sawflies, members of the order Hymenoptera, which includes ants and bees, are easily mistaken for caterpillars. How can you tell them apart?

Confusing duo:
A backswimmer,
below, is a predatory
aquatic true bug that
swims upside down.
(KIMBERLY R. FLEMING)
A water boatman,
right, is mostly a
scavenger that feeds on
or near the bottom of
ponds and pools.
(DICK TODD)

Sawfly larvae, like the one at top, have seven pairs of prolegs behind the three pairs of true legs by its head. In contrast, larvae of butterflies, like this Question Mark (middle), and moth caterpillars, like the "inchworm" (bottom), have fewer pairs of prolegs. (ERIC R. EATON)

Both caterpillars and sawfly larvae usually possess prolegs on the abdominal segments. These are not true legs, like the three pairs evident on the thorax, right behind the head. Instead, they are fleshy tubercles (knobs) equipped with gripping spines to help the insect hold onto a substrate and move efficiently. Caterpillars of moths and butterflies have at most five sets of prolegs, sometimes less in the case of "loopers" and "inchworms." Sawfly larvae have seven sets of prolegs. This may not be obvious, especially on the slimy-appearing "pear slug" sawflies.

— CHARLEY EISEMAN (HE/HIM)

Hi! I think I was interested in bugs and other living things as soon as I was old enough to walk around my yard. I remember picking up woolly bears, flipping rocks, and staring into the little pond down the hill from our house. My interest in bugs wasn't much encouraged during my formal education and didn't kick into high gear until a couple years after college, when someone gave me my first digital camera. Freed from having to be conservative about what I took pictures of, as I was with a film camera, I started photographing anything that caught my eye in nature, which ended up being a lot of bugs, their cocoons, egg sacs, leaf mines, and galls.

Charley Eiseman
(Julia Blyth)

I started collecting those mystery objects to see what would emerge from them. I accumulated enough to start wishing there was a book I could look them up in. I asked Mark Elbroch, author of *Bird Tracks & Sign* and *Mammal Tracks & Sign*, if he would ever write an "invertebrate tracks and sign" book, and he said it was up to me.

It hadn't occurred to me that writing a book is a way you can learn about something, as opposed to something you do when you are already an expert, but I accepted the challenge. Writing *Tracks & Sign of Insects and Other Invertebrates* taught me pretty much everything I know about bugs. Once I began the project, I discovered BugGuide, and having that online community to help me identify what I found was a big part of what made that project successful.

I tend to slowly prowl around whatever habitat I find myself in, with camera ready to get photos, and, of course, I spend time watching interesting behaviors without taking pictures. My main mode of bugwatching is to collect leaf mines, galls, and unknown larvae, especially sawflies, to rear to adults. A lot of these insects are poorly known, and my efforts often result in discovering new host associations and other new life-history information, as well as new species.

Much of the bugwatching I do is in my own yard, which, since my wife and I moved in to this property, we have transitioned from barren lawn into a diverse meadow, with fruit and nut trees, berry bushes, and vegetable gardens. Occasionally, we go on epic road trips in other bioregions to look for leaf miners that can't be found here in New England.

My greatest satisfaction as a bugwatcher comes from solving mysteries, seeing things I've never seen before, and learning about the interconnections among bugs, plants, and other organisms. Getting other people excited about insects they weren't even aware of before is also gratifying.

CHARLEY'S WEBSITE: https://bugtracks.wordpress.com/

CHAPTER 6

INSECT BEHAVIORS

ONE NEED NOT visit the Amazon or the Serengeti or turn on a television nature documentary to witness astonishing drama. Exploring your backyard can reveal life-and-death struggles, symbiotic relationships, and social cooperation among the insects. Common behaviors—such as feeding, courtship and mating, territoriality, nesting, swarming, mass emergences, and migration—offer clues to the identities of insects that cannot be communicated in most field guides and websites. Successful observation and photography of insects hinges on understanding and anticipating these behaviors.

— FEEDING

INSECTS EXCEL AT one thing in particular: eating. They are, in fact, our major competition for plants we grow for our own consumption, landscaping, and other uses. Predatory and parasitic insects eat each other, feed on the blood of vertebrate animals, or even prey upon small fish, reptiles, amphibians, birds, and mammals. Still other insects feed on fungi, decaying organic matter, fermenting sap from trees … Nearly any substance is fair game.

Damage: A pattern of defoliation can betray the presence of a plant-feeding insect, like this Bristly Roseslug Sawfly larva, *Cladius pectinicornis*. (ERIC R. EATON)

An easy way to find herbivorous insects is to look for damage to foliage and flowers. The insect responsible may be on the underside of the leaf or on a nearby twig. Nothing there? Try returning at night, as many insects feed under cover of darkness. Different insects leave different types of destruction in their wake. Some take entire chunks of leaves. Others are "skeletonizers" that leave a netlike pattern of unconsumed veins in the foliage. Turn yourself into a migrating warbler and glean every inch of a plant for traces of insects.

Predatory insects often wait in ambush on flowers or amid foliage. Even large species like mantises can be nearly invisible as they sit motionless while waiting for an oblivious insect to come near. Many assassin bugs use this technique, too. Active predators include robber flies that practice "hawking" like flycatcher birds: perch and scan the sky above, then dart out to snatch your prey. Such insects return to their perches to dine. Tiger beetles dash over beaches, dunes, and sunlit trails in pursuit of ground-dwelling prey.

Other predatory species favor gleaning, inspecting every inch of a plant or substrate for prey. Narrow-winged damselflies (family Coenagrionidae) are one example. They hover very close to a weed, slowly moving up and down the plant in search of another insect to pluck and eat.

Many insects feed in secret. They bore into dead trees or living ones. They burrow underground, like mole crickets and countless species of beetles and their grubs. They tunnel between layers of foliage (leaf miners) or inside of galls.

Insects also need to drink. Watching the edges of streams, lakes, ponds, and even mud puddles will let you see many insects coming to quench their thirst. Your birdbath or garden water feature may also attract them.

— COURTSHIP

The courtship rituals of insects can provide the bugwatcher with hours of spectacular entertainment and photo opportunities. The grand prize for Wildlife Photographer of the Year in 2022 went to an image of a frenzied ball of solitary cactus bees, *Diadasia* sp., taken in the Arizona desert by photographer Karine Aigner. While witnessing courtship and mating in insects is largely a matter of luck, there are ways to improve your chances. Patience is key. It can also be a matter of location, location, location. Walking on bare soil in spring or early summer, you increase the likelihood of seeing an aggregation of wasps or bees flying close to the ground. They are likely males waiting for virgin females to emerge. When one does, a frenzy will ensue, as in Karine Aigner's photo.

Among the more thrilling examples of courtship are certain robber flies, family Asilidae. Males may be ornamented with plumelike hairs on their

What a performance: A male robber fly, *Cyrtopogon* sp., hovers above a female. She is distracted by the prey she is eating. (SAMANTHA GALLAGHER)

bodies and/or legs. These accessories are displayed to the female while the male hovers in front of her. Robber flies in the genus *Cyrtopogon* are especially charismatic.

Males of other true flies may display patterned wings to the female, wave their legs, even dance to convince her he is the most genetically fit male to mate with. Pomace flies in the genus *Drosophila*, those tiny, familiar "fruit flies" in your kitchen, also have a courtship dance.

Male insects often engage in territorial behavior (see "Territoriality," page 112), even battles, as a prelude to or instead of true courtship. Likewise, males may gather in swarms or otherwise form leks to attract the attention of passing females. A lek is defined as a gathering of males, each of which may perform in competition with other males to attract mates. Think of a lek as a stage or arena performance of male animals.

The males of some dance flies, family Empididae, secure a prey item to present to the female, distracting her while he mates. A few species in the genus *Hilara* even wrap their "gift" in a balloon of silk, so that he has more time to mate while she unwraps it. Sometimes, a male gifts an empty balloon, as he can expend less energy by foregoing the hunting of prey. Similarly, hangingflies, which are not flies at all, but insects in the family Bittacidae, order Mecoptera (scorpionflies), will secure prey to offer the female.

Some insect males transfer protein, chemicals, or other vital resources to a female in hopes of eliciting her willingness to mate. These can include protein and/or minerals to aid in egg production or defensive chemicals that the female will pass along to her offspring.

— MATING

THE REPRODUCTIVE POTENTIAL of insects is legendary. You are very likely to observe insects in the act of procreation, though it may be difficult to recognize what is happening. Male and female grasshoppers, for example, contort their bodies in order to engage their reproductive organs, male atop the female, or side by side with her. Male dragonflies and damselflies grasp the female behind her head, using claspers at the tip of the male's abdomen. She in turn thrusts the tip of her abdomen forward to engage the male's genitalia, located at the base of his abdomen. The resulting coupling forms the "mating wheel," which is often appropriately heart-shaped.

Velvet ants and some related wasps with wingless females practice phoretic ("catching a ride") copulation. The winged male can be significantly larger than his wingless bride, and he may literally sweep her off her feet, carrying her to a perch to mate or mating with her while in flight.

Insects usually mate in one of two configurations: male atop female, oriented in the same direction, or "tail to tail," each pointed in the opposite direction. Some insects spend so much time in tandem that you seldom see them separately. Love bugs, *Plecia nearctica*, are a type of march fly. Couples can be together in mating position for up to two days. Many walkingstick insects, order Phasmida, also remain engaged for long periods. Male ambush bugs, assassin bugs in the genus *Phymata*, may ride females for days, but they are more often freeloading since the larger female can kill larger prey. Coupled insects can multitask. Male and female thread-waisted wasps, *Eremnophila aureonotata*, are often seen in pairs, flying perfectly well together, the female flying from flower to flower to drink nectar.

The infamous decapitation of males by female mantids is more myth than reality. In captivity, with no place to flee, he is more likely to become a meal. Still, it happens in the wild, too, and if the two are copulating beforehand, mating goes on despite partial cannibalism.

Postcoital behaviors include mate guarding, whereby the male continues to be in contact with the female or hovers closely while she deposits her eggs. This helps guarantee that he is the father of her progeny. This is common behavior in dragonflies and damselflies and other insects.

Hearts aflutter: Two mating bluet damselflies, *Enallagma* **sp., form a heart-shaped "mating wheel."** (SAMANTHA GALLAGHER)

— PARENTAL CARE

MANY INSECTS DEMONSTRATE some degree of parental care, beyond social bees, wasps, ants, and termites. Female earwigs of some species guard their brood of eggs and then young nymphs. Treehoppers in the family Membracidae do, too. Burying beetles in the genus *Nicrophorus* may work as

Protective mom: A female Oak Treehopper, *Platycotis vittata*, **guards her growing offspring from potential predators and parasitoid wasps.** (SAMANTHA GALLAGHER)

a couple or as single mothers to rear a small number of larval offspring atop "meatballs" created from small mammal, bird, or reptile carcasses. Single fathers exist, too: Male giant water bugs, family Belostomatidae, either permit the female to deposit her eggs on his back or guard batches of eggs that the female lays elsewhere.

— TERRITORIALITY

MALE INSECTS FREQUENTLY display territorial behavior. This often takes the form of stationing themselves for long periods on the ground, large rocks, twigs, or the foliage of plants. They are alert to both females and competing males, and will give chase or engage in ritualistic combat with rival males. Examples include many butterflies, some bumble bees, the Elm Sawfly, and cicada killer wasps. Scent marking of stems and other surfaces is another form of territoriality employed by the males of some solitary wasps and bees, like *Philanthus* and *Cerceris* wasps and *Anthophora* bees.

Resources used by females are another location where males attempt to assert dominance over other males. These include oviposition sites (places where females lay eggs), flowers where female bees gather pollen, the margins of water bodies where female wasps gather mud for nesting, and nesting sites themselves. Male sawyer beetles (*Monochamus* spp.) and other large longhorned beetles defend prime logs and dead or weakened trees coveted as oviposition sites by females. Dung beetle males will battle for the rights to a fresh mass of manure.

What a view: A hilltopping male branded skipper, *Hesperia* **sp., has no time to take in the scenery from a ridgeline in Colorado. He is on the lookout for receptive females and rival males instead.** (Eric R. Eaton)

— HILLTOPPING

Hike to the summit of a hill, plateau, butte, bluff, or other promontory, and you will probably encounter a number of insects stationed there or flying to and fro. These are males, using the highest point in the landscape as a place to intercept females. Many butterfly species, scarce under other conditions, will be found here. Males may clash with each other in territorial battles, but otherwise they are on the lookout for the opposite sex. Certain wasps, and flies, are also known to exhibit this behavior. A related behavior is landmark lekking, whereby males congregate at the top of isolated trees, shrubs, boulders, and other obvious objects.

— GROOMING

The survival of an insect depends on its ability to maintain the integrity of its physical body. Sensory organs and ambulatory appendages must be kept at their highest level of efficiency. Consequently, insects spend a great deal of time cleaning their antennae, eyes, wings, legs, and mouthparts. They need to perch to do this. Watch a sunlit wall of leaves at the edge of a field, prairie,

Contortionist: This male Drone Fly, *Eristalis tenax*, is cleaning his eyes with his hind legs. Do not try this at home. Insects groom constantly to keep their sensory and locomotory appendages in peak condition. (ERIC R. EATON)

or meadow, and in short order, you will notice many wasps, flies, and other insects alighting on the foliage to preen. That house fly on the kitchen counter, rubbing its "hands" together like a villain fiendishly plotting your demise? It is grooming.

— GRASSHOPPER BEHAVIOR

MANY GRASSHOPPERS, ESPECIALLY the ground-dwelling band-winged grasshoppers in the subfamily Oedipodinae, exhibit surprisingly complex behavior. Males engage in visual displays to each other and to females. Considering that these insects rely on near-perfect camouflage to avoid detection by predators, it seems counterintuitive that they would even have any bold colors for communication. They do, but the colors are concealed.

The inside of the femur of the hind leg is, in many species, banded in black and white, or even black and orange, red, or blue. The tibia joint on the hind leg may be bright yellow, red, orange, or blue. At rest, the insect hides these features. Should a competing male alight close by, the two begin dueling through femur tipping, flashing those bright colors by abruptly raising one or both hind legs to signal each other. Typically, a female is nearby, and they are attempting to assert dominance over a competing suitor. Females also use femur tipping to repel unwanted male attention.

Grasshopper behavior: Band-winged grasshoppers(top), with colorful
hindwings, make noisy crepitation flights to attract the attention of females.
On the ground, they communicate by femur-tipping, raising and lowering
their colorful hind legs (bottom). Slant-faced grasshoppers (right) will
avoid you by dodging behind grass stems. (Samantha Gallagher)

The hind wings of grasshoppers are broad and used in flight or to glide during leaps. In band-winged grasshoppers, these wings are often yellow, orange, red, pink, or blue, with a contrasting black band. In the Carolina Grasshopper, *Dissosteira carolina*, the entire wing is black, edged in cream or yellow. The males of this species perform a nearly silent, hovering flight over a prospective mate. In other band-winged grasshoppers, the males produce snapping and crackling sounds in flight. This is called *crepitation* and can be done at will. Exactly how the sound is produced remains mostly a mystery. The duration of crepitation flights varies greatly by species.

Slant-faced grasshoppers in the subfamily Gomphocerinae are arboreal in the sense that they live mostly off the ground among tall grasses. Their pointy heads and often vertically striped, linear bodies help them escape notice. They also dodge behind stems at your approach.

The courtship behavior of slant-faced grasshoppers is also auditory, but occurs when the insect is perched. Males generate sound by stridulation (see page 19), rubbing pegs on the inside of the femur of the hind leg against raised veins on the narrow front wings. The resulting "song" is typically an audible "zip-zip," repeated quickly, but with long intervals of silence, so as not to draw the attention of sharp-eared predators.

— THERMOREGULATION

BEING COLD-BLOODED, INSECTS need to warm themselves to become fully active. Many insects bask to do so. A basking insect typically adopts a specific posture, leaning to an angle that provides maximum exposure of their body to the incoming solar rays. Butterflies that appear "tipped over" are basking. This is the usual behavior for sulphurs and whites. Other butterflies bask with wings fully spread. Grasshoppers tilt themselves and drop the hind leg so that their abdomen receives all the warming sunlight. Solitary wasps will also lean or sprawl on the ground or a stone.

Once an insect reaches optimal temperature, any further increase risks overheating. Some go into hiding, sheltering from the hot sun. Others adjust their posture. Tiger beetles and grasshoppers prop themselves up on their legs to avoid scorching their bodies on the hot sand. They may even lift their feet alternately. Some dragonflies perform "obelisking" behavior, directing their abdomens straight up when perched, to minimize exposure to the midday sun. Other dragonflies seek shade and "droop," hanging vertically from a branch or stem.

Nocturnal and crepuscular insects cannot use the sun to warm themselves, so they have alternative strategies. The most commonly observed behavior is shivering, where moths appear to be idling their wings. The muscles inside the

Obelisking: A male Blue Dasher dragonfly, *Pachydiplax longipennis*, orients his abdomen vertically to minimize incoming solar radiation on his body in the heat of midday. (SAMANTHA GALLAGHER)

thorax, heavily insulated by a thick coating of scales, manufacture the body heat necessary for the insect to be active.

— GOBBETING

HAVE YOU EVER noticed a resting fly or bee apparently blowing bubbles? The creature is performing another type of thermoregulation through evaporative cooling. By exuding a droplet of water or regurgitated liquid, the insect exposes it to evaporation, which lowers the temperature of the fluid. When reingested, the droplet lowers the temperature of the insect's head and thorax. This accounts for the pulsing of bubbles from the tongue of a bee or the fleshy proboscis of a fly. The behavior also serves to concentrate nectar and other carbohydrates the insect needs for energy, while reducing the "cargo load" of liquid in the insect's crop, thereby improving flight efficiency.

— ROOSTING AND SLEEPING

THE MOST ENERGETIC of diurnal insects need rest. Solitary wasps and bees can be observed sleeping at dusk and at night. Despite their usual lonely nature, these insects may form loose or compact groups of several individuals, males especially, where they spend after-dark hours. Many, like mason wasps, scoliid wasps, and cuckoo wasps, grip a stem or twig with their jaws and/or legs and curl their bodies around it. *Ammophila* thread-waisted wasps adopt a 45°-angle posture, biting the stem but propping their bodies with their legs. When several are hanging from the same plant, they can be mistaken for slender, drooping seedpods. Mason wasps, cuckoo wasps, and cuckoo bees such as *Nomada* and *Xeromelecta* usually roost singly. Cuckoo bees grip the tip of a twig, leaf, or other plant part with their jaws and simply hang from their perch.

Steniolia sand wasps pack themselves into dense clusters on the tips of pine boughs and other foliage. Males of the Northern Blue Mud Dauber, *Chalybion*

Slumbering: Solitary wasps, like these thread-waisted wasps (left), *Ammophila* **sp., and solitary bees (right), like this** *Xeromelecta* **sp., spend the night and periods of unfavorable weather in awkward positions.** (SAMANTHA GALLAGHER)

californicum, gather in sheltering nooks on buildings and other human constructions, causing undo alarm given that they cannot sting. Male thynnid flower wasps, genus *Myzinum*, can festoon weedy plants by the dozens. Equipped with a "pseudo-sting" at the rear, they look menacing. Many male bees also spend the night in groups on weeds or alone inside folded flower blossoms. Solitary wasps that live in deserts will roost singly or in groups in foliage in the shade during the hottest hours. Dragonflies roost low in thick grasses, in the case of some larger darners, or hang vertically from perches higher in trees or shrubs. Look for them in the evening.

Edge habitat, where forest meets field, meadow, or prairie, is often the best place to look for slumbering insects. It is worth the search if only because they are much more tolerant of close approach at dusk or at night than they are during the day, allowing for perfect photo opportunities and close inspection for identification purposes.

— DIAPAUSE

INSECTS DO NOT hibernate in the same sense as vertebrate animals. Instead, they may weather unfavorable conditions, such as extreme temperatures or food scarcity, by undergoing diapause. Diapause is a cessation of growth, development, or reproduction, and a suppression of metabolic activity. It may be initiated by external cues such as changes in day length or be an obligatory change that occurs regardless of environmental stimuli. Diapause can occur at any stage in the life cycle, depending on the species. The Woolly Bear, larva of the Isabella Tiger Moth, *Pyrrharctia isabella*, goes through diapause in the caterpillar stage, usually sheltered in leaf litter or beneath a log or other debris. Female paper wasps and queen yellowjackets and hornets shelter as adults, wings curled beneath them, to overwinter. Many insects undergo diapause as pre-pupal larvae, within silken cocoons that provide added insulation against the elements.

— SWARMING

A *SWARM* IS a benign collection of insects, often for purposes of attracting mates or, in the case of honey bees, the departure of part of a colony from an existing hive. Dragonflies frequently assemble in aerial feeding swarms that may include more than one species. The verb *swarm* represents an action, such as an attack response, like what happens when you bump into a yellowjacket nest. Mating swarms can be awe-inspiring in their dimensions. Swarms of nonbiting midges, family Chironomidae, are usually mistaken for mosquitoes. Near lakes and other bodies of water, their swarms can appear as a living blizzard or waterspout.

All together now: Certain species of lady beetles, like these Convergent Lady Beetles, *Hippodamia convergens*, **spend unfavorable seasons in enormous congregations, often in mountainous terrain.** (ANTON SOROKIN)

— AGGREGATIONS

WHAT, AT FIRST glance, appears to be a swarm of insects may be something else. An aggregation of insects represents a gathering of individuals for purposes of seasonal diapause, sleeping, nesting, or other activity.

Common examples include boxelder bugs and lady beetles. Boxelder bugs, in the genus *Boisea*, gather in autumn at locations suitable for mass winter diapause. Those places are often the sides of buildings with a southern exposure. Ladybird beetles, especially in western North America, form spectacular aggregations in the cooler foothills and mountains during the height of summer, when their aphid prey is scarce in the scorching valleys.

Females of many species of solitary bees and wasps nest close together, giving the appearance of sociality. Each one has her own burrow but benefits from having many neighbors. With so many potential targets, crimes of parasitism, predation, and theft are less likely to hit *your* home.

Periodic eruption: Periodical cicadas in the genus *Magicicada* are unique to the United States. They emerge in vast numbers every 13 or 17 years, depending on the species. (ERIC R. EATON)

— MASS EMERGENCES

SWARMS CAN RESULT from mass emergences, such as certain mayflies along the shores of the Great Lakes. Historically, those mayflies are perhaps the most extreme example. Periodical cicadas may be the most celebrated. They are found exclusively in the eastern half of the United States and only emerge as adults every 13 or 17 years, depending on the species. Each "brood," of which there are 15 (2 others are extinct), known by a Roman numeral, such as Brood X, emerges synchronously in many different locations, over a period that peaks around Memorial Day weekend. It is a spectacle well worth witnessing, though it is best to be able to flee the deafening chorus of the males. Several websites, such as Cicada Mania, give maps and dates for each of the species and broods.

— MIGRATION AND MASS MOVEMENTS

MORE INSECT SPECIES undergo migrations than we previously understood. North American populations of the Monarch butterfly, *Danaus plexippus*, are known for their lengthy annual autumn migrations to coastal California, Mexico, and the Caribbean, but elsewhere in the world, the species stays put.

Recently, dragonflies, moths, planthoppers, the Large Milkweed Bug, and even some flower flies (family Syrphidae) have been discovered to migrate.

Not all species migrate consistently. Migration and mass movement may be triggered by food scarcity, for example. *Anabrus simplex*, the Mormon Cricket, is a type of wingless katydid that, in outbreak years, marches relentlessly over the landscape, devastating croplands in western North America. During population explosions, caterpillars of the White-lined Sphinx moth, *Hyles lineata*, may become nomadic in arid regions where food plants are widely spaced.

The larvae of some dark-winged fungus gnats, genus *Sciara* in the family Sciaridae, will undertake mass migrations, too, forming singular or branching columns of slimy maggots up to 3 feet long, even longer in some recorded cases. This behavior has earned them the name "snakeworm gnats," but scientists do not fully understand why it occurs.

Recording observations of perceived migrations of insects is of critical importance to science, as it may represent new knowledge or reflect new trends resulting from climate change.

— CHEMICAL DEFENSES

MANY INSECTS DEFEND themselves with chemicals, in a variety of ways. Perhaps the most famous are the bombardier beetles in the genus *Brachinus*, family Carabidae. They possess paired glands in the abdomen that funnel two chemical compounds into a reaction chamber where the combination of the two results in an explosive release of scalding-hot gas. The blast is accompanied by an audible pop, but it is actually a pulsating series of rapid-fire micro-explosions. The beetles occasionally fly to lights at night; handling them is not recommended.

Grasshoppers and sawfly larvae may regurgitate the contents of their crop when handled roughly. The "tobacco juice" of grasshoppers can leave a stain on human skin, but is not otherwise harmful.

— SELF-MUTILATION DEFENSES

BELIEVE IT OR not, insects may willfully bleed or willingly sacrifice body parts in the interest of self-preservation. Ladybird beetles employ autohemorrhaging, or "reflex bleeding," when assaulted. They ooze hemolymph from their body joints. It is sticky, stinky, and otherwise repellent to most potential predators, especially the ants that the beetles must contend with in order to feed on aphids and scale insects. The yellowish goo gums up the jaws of the ants. Fireflies and some leaf beetles autohemorrhage as well. Blister beetles (Meloidae) take it up a notch. Their blood contains a potent chemical called cantharidin that is toxic to most predators.

Sticky blood: Lady beetles, like this Streaked Lady Beetle, *Myzia pullata*, voluntarily ooze hemolymph (comparable to vertebrates' blood) in a phenomenon known as reflex bleeding. The blood coagulates and stinks, repelling and gumming up any would-be predators. (KIMBERLY R. FLEMING)

Have you ever held a grasshopper or cricket and had one of its legs come off in your hand? The insect is designed that way. The act of self-amputation, called autotomy, is equivalent to a lizard losing its tail on purpose. The hind legs of katydids and their kin have a weakened fracture plate at the joint of the trochanter and femur (think hip and thigh), allowing the leg to break free under sustained stress. It barely slows down the animal, and if it is a nymph, it can grow a replacement leg, albeit a smaller one, with its next molt.

— VISUAL DEFENSIVE BEHAVIORS

MANY INSECTS ADOPT characteristic, exaggerated attitudes in body language for purposes of self-defense. These displays can be bluffs or actual threats. They can be aggressive in nature or passive, to the point of feigning death.

Certain large moths, which usually present camouflage as their first line of defense, may, when prodded, expose large eyespot markings on the brightly

What big eyes you have: An Io Moth, *Automeris io*, raises her front wings to expose enormous eyespots on her hind wings. The tactic can startle or confuse an attacking predator. (ERIC R. EATON)

colored hind wings. This can be effective in startling a vertebrate predator into abandoning attempts to further attack the insect or, at least, deflect attention to the wing, which can absorb damage and still function.

Some caterpillars also sport eyespots that they can expose at will. Other caterpillars, when discovered, rear up and expose bright colors or thrash whiplike appendages menacingly.

— DEATH-FEIGNING BEHAVIOR

A NUMBER OF insects will, when fearful, "play possum" and do so convincingly. It can be aggravating to closely approach a peculiar weevil, leaf beetle, or other insect, only to have it fall off the leaf, twig, or other substrate it was perched on and be lost immediately in leaf litter, grass, or other ground cover. When feigning death, most insects retract their appendages, such that they will resemble seeds, pebbles, chips of bark, or other inanimate objects. They may remain motionless for periods long enough for most of us to lose patience. Insects that are already camouflaged and/or encrusted with debris, like ironclad beetles (*Phellopsis* spp.), hide beetles (*Trox* and *Omorgus*), and the Forked Fungus Beetle, *Bolitotherus cornutus*, are essentially invisible if they play dead. The warty leaf beetles, *Exema*, *Neochlamisus*, and related genera, are mimics of caterpillar frass (fecal pellets) and easily overlooked anyway. If you spot one, simply looking at it cockeyed will cause it to drop to the ground.

— DEBRIS-CARRYING BEHAVIOR

IF YOU SEE a small bit of lichen moving under its own power or an animated dust bunny, be assured you are not hallucinating. The larvae and nymphs of some insects disguise themselves by intentionally accumulating various materials on their bodies. This phenomenon is most frequently observed in aphidlions, the larvae of green lacewings. Several genera in the family Chrysopidae are known to do this, but it is apparently a choice of the individual insect. By decorating themselves in plant matter, bits of lichen, or even the empty exoskeletons of prey they have killed, lacewing larvae do not attract the attention of ants, which would otherwise kill them to protect the aphids that are dispensing honeydew to the ants.

Indoors, and in barns, sheds, and other buildings, nymphs of the Masked Hunter assassin bug, *Reduvius personatus*, mask their identities and predatory inclinations by coating themselves in lint and other debris. The Masked Hunter is known as a predator of bed bugs and their kin, but finding this insect does not mean those blood-sucking pests are present.

Disgusting defense: An Egg Plant Tortoise Beetle larva, *Gratiana pallidula*, makes a shield of its own dried fecal material that it carries like an umbrella on anal spines. (KIMBERLY R. FLEMING)

The caterpillars of bagworm moths, family Psychidae, weave bits of leaves or needles or twigs into the thick, silken case they use for protection and camouflage.

The larvae of tortoise beetles and their relatives, in the leaf beetle family Chrysomelidae, have a unique, if not disgusting, method for hiding and for repelling potential predators. They simply pile their fecal droppings onto the rear of their bodies. Depending on the species, the poop can be wet or dry. Sometimes it takes the form of an "umbrella," or delicate, symmetrical spirals. Case-making leaf beetles envelope their entire bodies in a capsule of their dry, rock-hard feces.

— PATROL PATTERNS OF MALE DRAGONFLIES

THE MALES OF many dragonfly species, especially those known as skimmers in the family Libellulidae, can be easily observed and photographed if one is familiar with their patterns of territorial behavior.

Blue Dasher, Twelve-spotted Skimmer, Widow Skimmer, Flame Skimmer, and several others regularly return to a preferred perch at the water's edge,

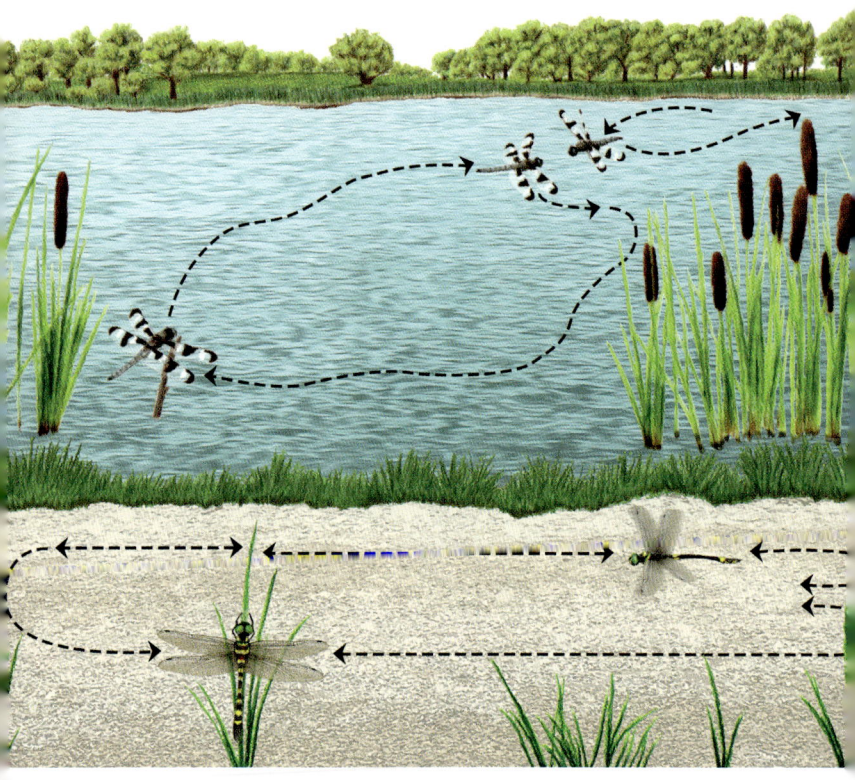

Flight plans: Males of many skimmer dragonflies, like the Twelve-spotted Skimmer, make patrol flights from a perch along the shore of a body of water, to look for females or chase off rival males. Others, like the river cruisers, *Macromia* spp., fly linear routes, parallel to rivers or streams, often some distance from shore. (Samantha Gallagher)

between forays to pursue females or chase away competing males. They may also fly regular loops around a pond or over a section of stream to defend their territories and/or locate available females. When a dragonfly leaves its perch, sneak closer so that when it returns you have an even better view.

River Cruisers are so named because the males typically fly linear paths along sections of streams or rivers, frequently along paths set back from the water's edge. They seldom deviate from their straight-line flights and may pass very close to a stationary human observer. At some point, they will perch, allowing for a better look if you have followed their flight.

— ANT BATTLES

ANT COLONIES ARE frequently in conflict with one another, and this can result in conspicuous behaviors easily observed. This may involve two different species or be a dispute between two colonies of the same species.

The Immigrant Pavement Ant, *Tetramorium immigrans*, is abundant in urban areas. Sometimes, two colonies come together in a spectacular territorial battle that manifests as a big pile of tiny ants on the edge of a sidewalk. The war may span many hours, but is largely ritualistic, with few, if any, casualties. It is apparently a measure of strength in numbers, but how the conflict is resolved, and when (withdrawal by one colony can take only 30 minutes), remains an unsolved mystery.

Amazon ants, genus *Polyergus*, are rarely seen, except when engaged in a "slave raid" on the nest of another ant species. Amazon ants send scouts from their nest to find colonies of certain species of *Formica* ants. When a scout returns with news of a target, large numbers of workers set out to invade the *Formica* nest, kidnapping the pupae of their adversary and killing the workers attempting to defend their vulnerable brood. These raids take the form of long lines of red ants going from their own nest to the colony they are plundering and back again. Once inside the nests of their captors, the pupae finish metamorphosis to become servants of the Amazons. Ironically, some species of *Formica* ants are facultative raiders of other species of *Formica*, meaning that they can survive independently, but will raid the nests of other species opportunistically. *Polyergus* Amazon ants are obligatory slave-makers, unable to survive without the assistance of their captives.

— BUGS ON BUGS

PHORESY IS THE scientific term for hitchhiking, and you may sometimes observe insects riding on unrelated insects. The most commonly seen example involves a tiny arachnid called a pseudoscorpion. As their name implies, pseudoscorpions resemble diminutive scorpions, minus the tail. They rarely exceed 5 millimeters, are exceptionally flat, and live under bark and in leaf litter, caves, and similar niches. Being wingless, pseudoscorpions rely on flying insects to ferry them to new locations. Favorite carriers include longhorned wood-boring beetles and braconid wasps. The little arachnids will sneak under the elytra (wing covers) of larger beetles, but clamp their claws onto the legs of other insects. They can be mistaken for large mites at first glance.

Look for wasps or bees that appear to have a seedlike object jutting out from between segments. That is a twisted-wing parasite, in the order Strepsiptera. The adult host usually picks up the active larvae of the parasite (referred to as a stylopid) from a flower it visits. The larvae are carried back to the nest of the wasp, where they attack the host egg or larva, feeding on its blood. Host and parasite develop in synchrony, and pupation of the stylopid occurs between the adult host's abdominal segments. Adult female stylopids stay put, releasing pheromones that attract the winged males. Eggs hatch inside the female's body, and the larvae exit her, and the host, when the host wasp or bee lands on a flower.

The active, first-instar larvae of blister beetles and wedge-shaped beetles are called triungulins. Like stylopids, they wait on flowers for a bee or wasp to alight, then climb aboard. Carried to the nest, the triungulin becomes a parasitoid (a parasite that invariably kills its host) of the host insect's larval offspring.

— INSECT PARASITES AND FUNGAL PATHOGENS

Bugs are plagued by other lethal organisms, especially mites and fungi. Mites on insects can simply be hitching a ride to a new location or food resource, or be feeding as parasites on the host insect. Mites that travel on carrion beetles, dung beetles, and fungus beetles are typically getting transport to a carcass or dung pat where they will feed on fly eggs. Bright red mites on insects are usually parasitic, feeding on the host's hemolymph (blood).

A dead, bloated, fuzzy fly hanging from the underside of a leaf or other vegetation is likely the victim of a pathogenic fungus peculiar to flies: *Entomophthora muscae*. The fungus takes control of the infected insect's behavior, causing "summit disease," whereby the host ascends to a high point and adopts a posture that will guarantee the fungal spores will burst through seams in its exoskeleton to rain down on other flies.

Mite-y problem? Red, stationary mites on insects and arachnids are usually parasitic, like this one (left) sucking hemolymph (insect "blood") from an aphid. (Kimberly R. Fleming) The brown mites roving over this ground beetle (right) are merely using the beetle as transportation to a new location. (Anton Sorokin)

Fungus-riddled: A fly-specific fungus, *Entomophthora muscae,* **has taken down this fly (left), causing it to perch, then burst, raining spores on any other flies passing beneath it. An unidentified fungus has overtaken and killed a Carolina Leafroller,** *Camptonotus carolinensis* **(right).** (Eric R. Eaton)

A similar fungus, *Entomophaga grylli*, afflicts grasshoppers, which will exhibit the same climbing behavior, embracing a plant stem in a death grip before exploding with fungal spores. The poor grasshopper cadaver may literally lose its head as a result.

Stranger still are horsehair worms, which are in their own phylum, Nematomorpha. These creatures lay millions of eggs in fresh water, where the ova are consumed by aquatic animals. Only those aquatic insects that can leave the water are appropriate first hosts. The worms form cysts in the host's digestive tract. When the insect leaves the water and dies on land, it must then be consumed by a grasshopper, cricket, mantis, or related insect for the worm's life cycle to continue. Inside this second host, the worm completes its development, eventually taking up most of the abdomen. It releases chemicals that instruct the normally terrestrial host to submerge itself in water, whereupon the adult worm, up to 4 inches in length, emerges.

— *JOSEPH SAUNDERS/JD MONROE (HE/THEY)*

Hi! I have always had empathy for species that are vulnerable and suffer a lack of appreciation by humans. I gravitated first to reptiles and amphibians, as photography subjects. Using a manual wheelchair has limited my pursuit of those animals. After acquiring a macro lens to improve my herpetology photography, I realized I could fill the frame with a medium-sized insect and "bugs" were much easier to find.

I found that even staring at one shrub long enough revealed a number of different insect species. You do not need a lot of space to have a productive day of observing and photographing. Persistent observation will also train your eye to notice when certain insects are more apt to be cooperative subjects.

If I am on my own, looking for wildlife of any kind (I am a birder, too), I prefer spaces where man-made structures meet the wild, like lake parks or city parks that are not heavily manicured, but are still accessible. I often visit park benches with trees overhanging them, making insect subjects easier to reach. I am also fortunate to have friends who regularly set up lights and sheets to attract moths and other invertebrates at night.

Joseph Saunders.
(Helen Rodriguez)

Still, finding accessible places, here in Oklahoma anyway, that are not drenched in pesticides, herbicides, and other chemicals that diminish insect diversity is a real challenge. Then there is the racial profiling that all Black people face. I have been chased out of locations by civilians, told to go back to where I came from, and had people call the cops on me. I have been accused of committing crimes that I am physically incapable of doing and been otherwise harassed for being a Black man in rural areas seeking wildlife subjects to photograph. These are obstacles that cannot be overcome without banishing racial oppression and ableism.

On the positive side, I derive the most satisfaction when people engage with me over my photographic work. I see myself as more of an artist than a scientist, but people tell me my images help them perceive invertebrates, reptiles, and amphibians as nonthreatening, even inviting. Photography is my way of advocating for other life forms.

It all comes back to my identity as a Black disabled man, often thought to have as little value to society as most people perceive bugs to have. It's a sort of kinship with them (herps, too). I know how valuable we are to our respective environments and that we are not treated accordingly. One truth that snakes taught me is that another person or animal does not have to accept you in order to have value.

It is not enough to stop injustice in pursuit of a just society. It is absolutely necessary to purposefully create an affirming and nurturing environment for all people and all living things to thrive in. That includes celebrating characteristics of a person or demographic that make them explicitly different.

Joseph's website: https://paraherpetologica.com/

CHAPTER 7

SOCIAL BUGWATCHING

Traditionally, bugwatching has been a solitary pursuit, and many prefer it that way, but opportunities to watch and learn about insects with others have increased exponentially in recent years. There are in-person and online communities to connect with, workshops and webinars to attend, insect-centered events, festivals, and even a fledgling insect ecotourism industry.

— COMMUNITY SCIENCE

Formerly known as citizen science, community science has done more to erase the stereotypes of nature enthusiasts than any other recent social development. It has done so through the relative anonymity of the internet, allowing introverts to participate to whatever degree they are comfortable with. Community science also gives purpose to nature observation and recording. Community science still has far to go in making minority

Informal expedition: A group from the Mile High Bug Club, in Colorado, explores a wetland. Such outings allow participants to learn from each other and help each other spot elusive insects. (Heidi Eaton)

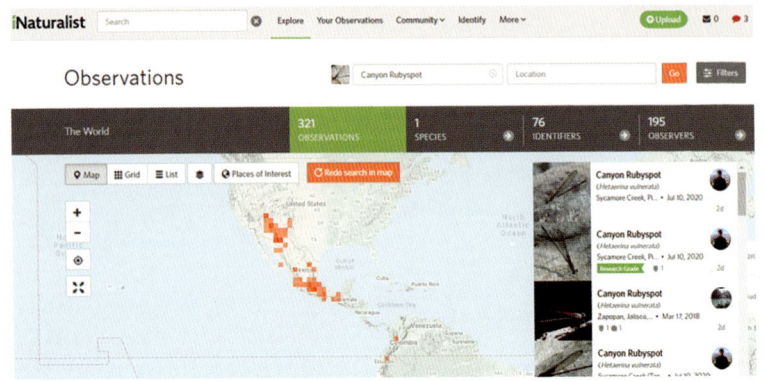

Online exploring: Websites like iNaturalist are a way for even the casual bugwatcher to contribute to science and to find out what other naturalists are seeing and where they are seeing it. (NANCY MCINTYRE)

demographic categories welcome and appreciated, but it is making consistent strides in that direction.

iNaturalist, with a website and app, is currently the leading platform for global community science. Anyone can register for free, though donations to this nonprofit are encouraged. It is user-friendly in navigation, with easy uploading of images or audio files. "Projects" on the site are additional places you can deposit observations. Many government agencies, schools, and organizations use iNaturalist for documenting events like bioblitzes (communal efforts to record as many species within a location and time period as possible). iNaturalist also has non-English language accessibility, and projects for LGBTQ and ADHD communities.

BugGuide predates iNaturalist, but covers only terrestrial North American arthropods. Membership is free, the user interface relatively straightforward. It was the creation of a single individual and became so popular that it quickly exceeded their server space. Fortunately, the entomology department at Iowa State University agreed to inherit BugGuide, and it continues to be maintained and periodically upgraded there.

Specific community science endeavors proliferate constantly, with varying degrees of endurance. Many are run by museums or other nonprofit organizations. Most are not heavily publicized and take a bit of searching to locate. Sci-starter is something of a clearinghouse, where users search for community science projects of interest to them. Popular insect-related projects include the Lost Ladybug Project, Firefly Watch, Monarch Watch, Bumble Bee Atlas, and the Great Sunflower Project (pollinators).

— COMMUNITIES OF INSECT-WATCHERS, REAL AND VIRTUAL

ONE NEED NOT look far for bugwatching friends online. There are interest groups on Facebook, Bluesky, and LinkedIn, plus Instagram, Pinterest, Tik Tok, and Reddit. By now, there may be new outlets. Finding in-person organizations where you live may be more difficult.

There are "bug clubs" and other organizations, but they suffer from a lack of publicity and frequently blink out of existence. Most of them experience initial increases in membership, but then reach a plateau that continues for years or decades. Most organizations focused on arthropods are for professional scientists. A slightly smaller number are open to both professionals and amateurs. Many amateur groups are dedicated to the trade in exotic, live arthropods such as tarantulas, mantids, and cockroaches. Others are limited in scope to honey bees, or native butterflies, dragonflies, moths, and sometimes beetles; at least they present as being less enthusiastic about other kinds of insects.

General bugwatching clubs include the Lorquin Entomological Society in southern California, which counts professionals and amateurs among its membership. The Webster Groves Nature Study Society, located in Missouri, has an entomology subgroup. The Mile High Bug Club, headquartered in Colorado Springs, Colorado, participates in a variety of educational, scientific, and conservation endeavors. 4-H remains a strong leader for youth entomology, and you might contact the Cooperative Extension Service agent(s) in your county to see if there are other "bug" groups in your area. If not, start your own. You may be surprised by the interest you receive from closet bugwatchers.

Ideally, the United States and Canada can look toward Europe for proven models of networking for bugwatchers. The Tanyptera Project, funded by the Tanyptera Trust, is headquartered at the Liverpool Museum in the United Kingdom. This conservation project was initially set to run from 2017 to 2024, to unite community scientists for documentation of terrestrial invertebrate fauna of the northwestern United Kingdom. It loans equipment, provides training, digitizes older literature, and otherwise invests in its volunteers for improved results. Tanyptera Project is such a success that funding has been extended through 2029.

— A WORD ABOUT INCLUSION AND ACCESSIBILITY

NATURE STUDY AND recreation, including bugwatching, is for everyone, or it should be. Early progress in entomology and other biological sciences was achieved by exploiting the labor of women and minorities under patriarchy and colonialism. We are still working to disassemble the harmful power structures that inhibit access to science in the present day, in order to restore trust in the sciences and encourage participation. Many women, non-binary people and those who are Black, Latino, Latina, Indigenous, and otherwise non-White, non-cis, or non-straight are leaders in entomology and science communication. Positive change can be accelerated by promoting minorities into leadership roles at every opportunity.

Efforts are underway to fund scholarships, field experiences, and other avenues for marginalized demographics to participate in bugwatching and/or entomology as a career path. Links to these organizations are listed at the back of this book. Please donate to them to keep the momentum going.

Meanwhile, leaders of nature walks and various outdoor recreation programs assume a great deal in their methodology of observing insects. We seldom, if ever, consider those who are visually impaired, for example. Other potential participants may not be fully ambulatory. Those who are neurodivergent may become overstimulated. What are we doing to make sure disabilities are not a barrier to exploring the natural world? The answer, until recently, is precious little. The organization Birdability came into existence in 2021, but is already leading the charge in the accommodation of those with substantial challenges to field outings.

The costs of equipment, admissions into parks, nature centers, refuges, and other properties and attractions, and membership in various organizations can be a barrier to many. We need to eliminate that kind of financial gatekeeping. Can people even *get* to parks and natural areas without a personal vehicle?

Demonstrating your commitment to inclusiveness can take many forms. Make a habit of stating your own preferred pronouns, such as he/him, she/her, or they/them, when introducing yourself. Perhaps learn some American Sign Language. Maybe learn the names of different insects in other languages. Acknowledging aloud that your outing is taking place on Indigenous land, where you represent as a settler or colonist, can be beneficial and set a standard for others to follow. Offer to pay admission for others to the park you frequent.

Lastly, creating a nonhostile social environment for children, women, transgender persons, and non-binary people is of paramount importance. Sexual harassment cannot be tolerated, nor can bullying and other forms of physical, emotional, and intellectual abuse. It is best not to assume another person's degree of knowledge and experience, and to avoid condescending remarks. Thank you for recognizing the need for humility and empathy when listening to the individual needs of others.

We can broaden the inclusivity of bug clubs, and improve their safety, in a variety of ways. Making it clear that your group is a safe place for everyone at all times is a step in the right direction. Promoting Indigenous, Black, Latinx, and other persons of color to positions of leadership and authority expands the realm of ideas and experiences for everyone. Actively soliciting input from the LGBTQ, neurodivergent, and other minority communities, and listening carefully to their needs and requests, will improve the quality of your organization, too. Anticipating the needs of the handicapped and neurodivergent helps relieve them of the burden of asking for accommodations.

Is your meeting venue or field-trip destination handicapped accessible? Is it in a location comfortable for non-White participants? Offering women-only programs and outings can increase participation, especially if you embrace transgender persons in those events.

It is not only the neurodivergent and autistic who appreciate small group sizes, the option to leave an activity at any time, or having an established plan for an outing before it starts (exactly where and when to meet, with map included, and when the event will end). Being mindful of textures that cause discomfort, like tall grass and wet clothing, is also helpful.

— WORKSHOPS AND WEBINARS

Informal in-person and virtual courses are being offered with increasing frequency. The cost is usually reasonable, and sometimes scholarships and other financial aid are available.

The Southwest Research Station in Portal, Arizona, USA, operated by the American Museum of Natural History, has for many years been offering separate, weeklong residential courses in the identification of ants, beetles, moths, bees, flies, and other insects, led by world authorities on those insects. Collecting and preserving specimens is part of these intensive workshops. The programs attract many graduate students from all over the world, but it is worth applying even if you consider yourself a relative amateur.

The Cincinnati Museum Center, together with the Nature Conservancy, runs the Eulett Center at the Edge of Appalachia Preserve in Adams County, Ohio. Each year they offer on-site mini-courses in the identification of various organisms, including insects. These programs usually involve in-the-field and in-the-lab observation of living specimens.

Since 2021, an annual online WaspID course, sponsored by Pennsylvania State University, has attracted a diversity of students from around the globe. The instructors are exceedingly patient and welcoming, and the students help each other, too. Identification of wasps is accomplished using specimens from existing institutional collections, imaged live through microscopes. It is well worth the experience, and a great model for future workshops on other kinds of insects.

Master Naturalist programs, an offshoot of Master Gardeners, exist all across the United States. While "master" is understandably off-putting for people of color, the curricula are worthy of exploration.

A nearly infinite number of organizations offer webinars and lectures about insects. These are usually offered live, sometimes streamed on social media, and often archived on YouTube or on an organization's own website. The more "plugged in" you are to entomology and natural history networks, the more likely you are to receive notice of such events.

— EVENTS FOR BUGWATCHERS

Beyond clubs and organizations, there are events where you can meet other bugwatchers in person. National Moth Week, Fourth of July butterfly counts, the City Nature Challenge, and other community science projects have blossomed into much-anticipated events.

Bioblitzes provide opportunities to engage with scientists in the field, and to "cross-pollinate" with botanists, mammologists, and other biologists. Bioblitzes are timed events at specific locations where the goal is for teams to document as many species of plant and animal life as they can. Some bioblitzes are more specific, targeting dragonflies, for example, or bumble bees. The

Scenic science: A bioblitz targeting dragonflies and damselflies along the Gila River in southwestern New Mexico. It doesn't get much better than wilderness landscapes and comradery with a purpose. (Eric R. Eaton)

definition of a bioblitz has expanded recently to include indefinite (no time limit) documentation of organisms at a specific location, to accommodate the ongoing recording of trends in species diversity and abundance over time.

The City Nature Challenge, which takes place at the end of April each year, began as a friendly battle between the cities of Los Angeles and San Francisco to see which city could record the most species of flora and fauna. It is now an international celebration and community science event.

National Moth Week, now an international project, occurs during the last full week of July each year. There are public events, typically at parks or nature centers, where black lights are hung in front of white sheets at night in hopes of attracting nocturnal moths. Private events may range from a neighborhood gathering to simply turning on your porch light to see what flies in.

Fourth of July butterfly counts are held over an indefinite period in summer in the United States and Canada. They are conducted under the auspices of the North American Butterfly Association, though they began as a project of the Xerces Society in the 1970s. A "hub" is selected for each count, with a radius of 7.5 miles. Participants form teams, with each team assigned a section of the count circle.

Among the more innovative online offerings the author has come across is a weekly insect "sketch-along" with entomologist Trisha Nichols on YouTube. The participants select an insect specimen from a small set that Trisha has pulled from her collection. Putting it under a microscope and sharing her screen, she invites everyone to sketch the specimen along with her. It is a creative and exciting way to learn insect anatomy, biology, and scientific illustration.

— FESTIVALS

THERE ARE SO many insect-themed festivals across the United States that the Insect Festival Working Group was formed to facilitate best practices for events that are time-limited and recurring. Indeed, most "bug fests" are yearly in nature.

The annual Bug Fair, at the Natural History Museum of Los Angeles County in southern California, is so popular that it regularly breaks attendance records for the museum. This happens the third weekend in May each year. Vendors of preserved specimens, live insects and arachnids, artwork, books, clothing, and other merchandise line the halls, along with booths for educational and government organizations. Speakers are invited to present, chefs cook up edible insects, and a house band, dressed in bug costumes, entertains. It is an exceptionally fun event.

An outstanding field-centered festival is Mothapalooza, held annually in south-central Ohio. The first one was in 2013, and it has been accelerating in

Scenes from a festival: The annual Bug Fair at the Natural History Museum of Los Angeles County draws vendors of living and preserved insect and arachnid specimens, plus artists, booksellers, and educational organizations. It draws about 2,000 guests over two days. (Chris Alderman)

popularity ever since. There are day trips, a banquet, speakers, and vendors, among other offerings.

One of the most unique and enduring bug celebrations is the annual Insect Fear Film Festival at the University of Illinois, Urbana-Champaign. It was founded in 1984 by Dr. May Berenbaum, a recipient of the National Medal of Science in 2014. For one day in February, Berenbaum and her students use bug-themed science fiction, fantasy, and horror classics as a way to educate the audience about actual insect biology. The event is hugely popular and has drawn national media attention.

— INSECT ECOTOURISM

ADVENTURING FOR INSECTS around the world has, so far, lagged behind birding-focused tourism, but these types of programs, especially butterfly-centered packages, are gaining momentum. Museums and nature centers, resorts, and summer camps are also beginning to recognize the allure of hunting insects.

Butterfly excursions in Costa Rica and Taiwan draw around 500,000 tourists per year, and that statistic was cited in a research paper 14 years ago. The Valley of Butterflies on the island of Rhodes, Greece, ironically features large congregations of a diurnal tiger moth, *Euplagia quadripunctaria rhodosensis*.

Bioluminescent insects seem to have an even more captivating power than butterflies. Synchronous firefly phenomena have been a draw for tourists in Malaysia for decades. When a different species, *Photinus carolinus*, was identified in the Great Smoky Mountains of the United States, it, too, became a destination spectacle; those hoping to attend apply months in advance for vehicle passes that are awarded by lottery. Meanwhile, in certain caves in New Zealand and Australia, glowworms (in this case, the predatory larvae of fungus gnats) enchant travelers seeking an experience to be found nowhere else.

Rainforest hike: There is nothing like experiencing a new natural realm. Travel if you can; you will not regret the experience.
(KELLI WALKER)

Lightning in a beetle: Firefly watching is becoming as big a nocturnal activity as stargazing. The beetle family Lampyridae has species all across the United States, though not all of them are bioluminescent in the adult stage. (ANN OLIVER FOR AUDUBON SOCIETY OF OHIO)

Make sure you don't bring back bed bugs from your vacation! Inspect your lodging accommodations thoroughly before settling in. (ERIC R. EATON)

Bed-and-breakfast owners are beginning to see the potential of touting natural history experiences, including bugwatching. Ask proprietors for more information. Arizona, Texas, and Florida are among the destination states most likely to offer nature-related options.

— *BRYNA B. (SHE/HER)*

An ecosystem is a narrative, an active narrative, that is playing out right in front of us.

— BRYNA B.

AS A TODDLER in Canada, I only ever wanted to look at the pictures in old field guides in my mom's music room. While she taught piano, I sat on the floor and flipped through the plates of bugs, trees, and reptiles of North America. After getting my first big-kid bike, I would head straight to the swamps by the Ottawa River and spend the afternoon on my tummy looking at all the life there was in the swamp, trying to ID it.

When I was young, I lived for dip-netting aquatic insects. Now I mostly go for nature walks and sometimes stay up all night in front of a mercury-vapor bulb looking at the many moths and other insects that are drawn to it. While vernal

Bryna B. (BRYNA B.)

ponds are my first love, I live in the tropics now and primarily bugwatch on my property at the Monteverde Butterfly Gardens in Costa Rica.

Bugwatching is not always idyllic. I've always had to deal with men approaching me, feigning interest in what I'm doing in order to get my attention. I have had several close calls while alone in the woods. I find it unfortunate that I sometimes feel uncomfortable being out in nature by myself, having to ask male colleagues or friends to come with me in order to avoid unwanted attention. As a woman working in ecotourism, I have had negative run-ins with male guides trying to undermine or argue with my identifications or try to delegitimize my thoughts and comments.

When someone tries to make me smaller in front of clients or other people, I stick to what I know and speak directly to the other people while ignoring the person trying to bring me down. I also report such behavior to the agencies that these guides work for. Women are incredible naturalist guides.

We need diversification and representation in guiding, and in science in general, so that we can include and connect with more people. We need everyone to get the chance to see and know that we have the coolest planet. On television and on guided naturalist tours, there is a clear bias toward White men. White, straight-presenting men cannot be the only ones permitted to represent our planet.

Being able to connect with someone who was at first clear about their dislike for insects, and show them that the beautiful diversity on our planet is what makes it so special, is what I live for. I spend every day advocating for the unloved by giving tours and training scientists at our facility. We get other interested people to share our world with others through our internship program. We work hard to create an inclusive and safe space for everyone so that we get a diverse group of interns who go out to be the voice for the incredible animals on our planet.

The thing that I'm most proud of in my career as a science translator is having trained more than 300 interns from around the world to share insects with all kinds of people in a fun and engaging way. I'm also proud of running a facility recognized for its impeccable science translation for over 12 years. We receive over 14,000 visitors a year.

Bryna's website: https://www.brynascience.com/

CHAPTER 8

ADVANCED BUGWATCHING AND ENTOMOLOGY

Casual observations of living insects in the field have been considered by the scientific community to be anecdotal, not rigorous, academic-based studies. While, for decades, biologists in European nations have embraced the work of amateurs as an important complement to professional science, their colleagues in the United States have reluctantly recognized the contributions of nonscientists, and only in the past decade or so. How can the bugwatcher become an even more valuable participant in the scientific process?

— COLLECTING SPECIMENS

Most field guides and other popular entomology books include a chapter on how to make an insect collection, but no one with an interest in insects

Why so many? Large collections of pinned insect specimens, like the holdings at the Denver Museum of Nature and Science, may appear to be overkill, but each specimen usually represents a different geographic location, date, and/or population. They represent history that can inform the future of biodiversity conservation. (Heidi Eaton)

Curated cabinets: The author explores a museum collection. Climate-controlled, pest-proof facilities like this ensure that scientists will have specimens for research for decades, if not centuries, to come. (Heidi Eaton)

is obligated to do so. It is, however, important to understand why there is the need for scientists to do that. Increasingly, our understanding of the relationships between species, and other taxa, over time, hinges on analysis of mitochondrial DNA and other types of genetic material that can only be extracted from preserved specimens. Likewise, analysis of the amounts of certain elements (as in the periodic table) in specimens can illuminate how changes in climate and environment have affected a given species. There is simply a great deal that can be learned from new and existing collections.

Building a personal insect collection requires a substantial financial investment to be done properly. This is cost prohibitive for many, and it also requires physical space that may not be available in a home or apartment. There is also little return on investment when one eventually donates their collection, assuming we will collectively still value university and museum repositories in the future.

The bugwatcher, if so inclined, can help entomologists by plugging into academic circles via email listservs and social media. Watch for requests for specimens from your particular geographic region, and volunteer to collect some, closely following the instructions for how to do so. Agree to allow the placement of traps on your property. This is exceptionally helpful in monitoring the spread and abundance of nonnative species, and in helping identify species new to science. One community science project in Los Angeles, California, where volunteers agreed to have traps placed in their yards, yielded over 30 species of tiny scuttle flies, family Phoridae, that were new to science.

— ADVANCED IDENTIFICATION OF INSECTS

IDENTIFYING MOST INSECTS to species, or even genus, necessitates a deceased, properly prepared specimen, a quality binocular dissecting microscope, and an excellent illuminator. Additionally, it requires documents known as binary keys.

Microscopes of even adequate standards usually cost over $1,000, and that includes refurbished older models, which are generally the least expensive. The

Students of entomology: Michelle Sloan Bos (left) and Marci Hess (right) have taken their pursuit of bugwatching to the next level in their home laboratories. (BROOKE BOS AND ANJA NARDI, RESPECTIVELY)

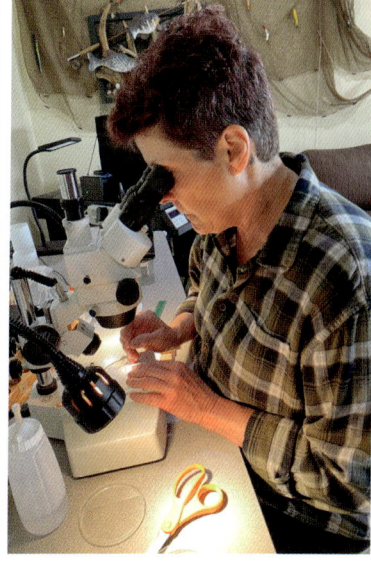

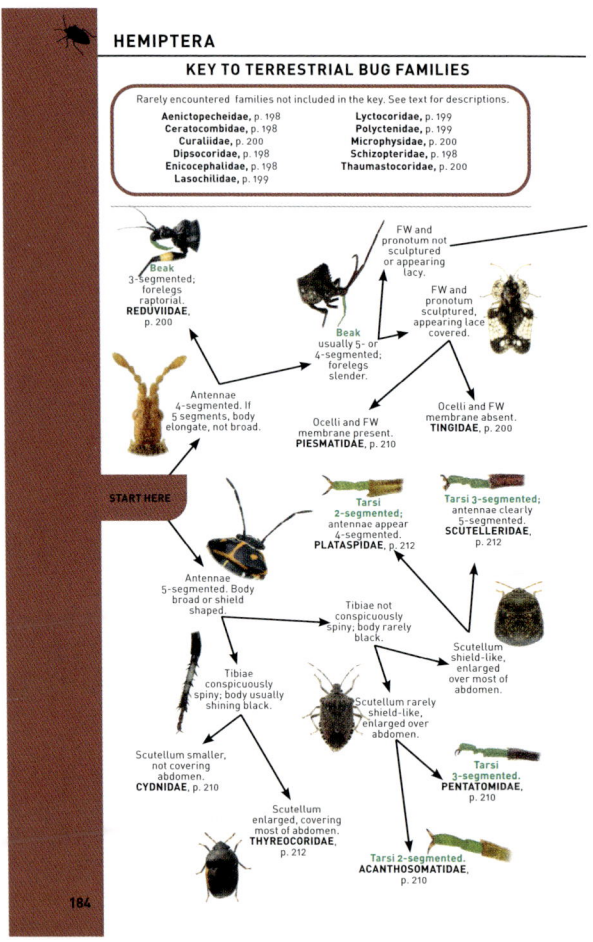

HEMIPTERA

KEY TO TERRESTRIAL BUG FAMILIES

Rarely encountered families not included in the key. See text for descriptions.

Aenictopecheidae, p. 198	**Lyctocoridae**, p. 199
Ceratocombidae, p. 198	**Polyctenidae**, p. 199
Curaliidae, p. 200	**Microphysidae**, p. 200
Dipsocoridae, p. 198	**Schizopteridae**, p. 198
Enicocephalidae, p. 198	**Thaumastocoridae**, p. 200
Lasochilidae, p. 199	

Beak
3-segmented;
forelegs
raptorial.
REDUVIIDAE,
p. 200

FW and
pronotum not
sculptured
or appearing
lacy.

FW and
pronotum
sculptured,
appearing lace
covered.

Beak
usually 5- or
4-segmented;
forelegs
slender.

Antennae
4-segmented. If
5 segments, body
elongate, not broad.

Ocelli and FW
membrane present.
PIESMATIDAE, p. 210

Ocelli and FW
membrane absent.
TINGIDAE, p. 200

START HERE

**Tarsi
2-segmented**;
antennae appear
4-segmented.
PLATASPIDAE, p. 212

Tarsi 3-segmented;
antennae clearly
5-segmented.
SCUTELLERIDAE,
p. 212

Antennae
5-segmented. Body
broad or shield
shaped.

Tibiae not
conspicuously
spiny; body rarely
black.

Scutellum
shield-like,
enlarged
over most of
abdomen.

Tibiae
conspicuously
spiny; body usually
shining black.

Scutellum rarely
shield-like,
enlarged over
abdomen.

**Tarsi
3-segmented**.
PENTATOMIDAE,
p. 210

Scutellum smaller,
not covering
abdomen.
CYDNIDAE, p. 210

Scutellum
enlarged, covering
most of abdomen.
THYREOCORIDAE,
p. 212

Tarsi 2-segmented.
ACANTHOSOMATIDAE,
p. 210

184

Keys to the kingdom: Identification of insects is made easier with a specimen in hand and access to a key in a book or online. (REPRODUCED FROM ABBOTT, J. C., AND K. ABBOTT, *INSECTS OF NORTH AMERICA*. PRINCETON UNIVERSITY PRESS, 2023)

light technology for illuminators continues to be upgraded, from bulbs to optical fiber to LED and so on. It may be worth investigating to see if a local nature center, college, or even high school would be willing to let you have access to one of their microscopes.

Keys demand an advanced comprehension of the terminology for insect anatomy, which is frequently different from one order of insects to the next.

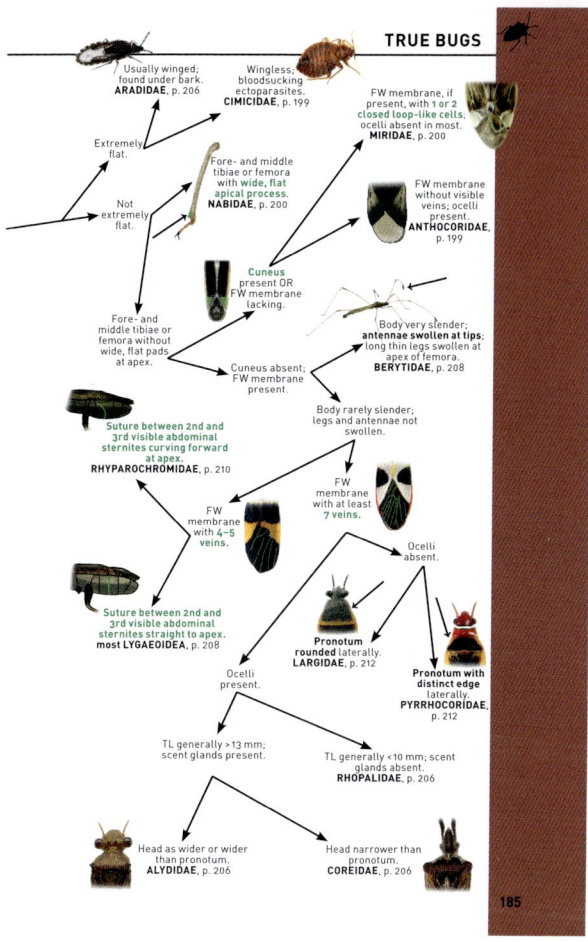

TRUE BUGS

Usually winged; found under bark.
ARADIDAE, p. 206

Wingless; bloodsucking ectoparasites.
CIMICIDAE, p. 199

FW membrane, if present, with 1 or 2 closed loop-like cells; ocelli absent in most.
MIRIDAE, p. 200

Extremely flat.

Fore- and middle tibiae or femora with **wide, flat apical process.**
NABIDAE, p. 200

Not extremely flat.

FW membrane without visible veins; ocelli present.
ANTHOCORIDAE, p. 199

Cuneus present OR FW membrane lacking.

Fore- and middle tibiae or femora without wide, flat pads at apex.

Cuneus absent; FW membrane present.

Body very slender; **antennae swollen at tips**; long thin legs swollen at apex of femora.
BERYTIDAE, p. 208

Body rarely slender; legs and antennae not swollen.

Suture between 2nd and 3rd visible abdominal sternites curving forward at apex.
RHYPAROCHROMIDAE, p. 210

FW membrane with 4–5 veins.

FW membrane with at least **7 veins.**

Ocelli absent.

Suture between 2nd and 3rd visible abdominal sternites straight to apex.
most **LYGAEOIDEA**, p. 208

Pronotum rounded laterally.
LARGIDAE, p. 212

Pronotum with distinct edge laterally.
PYRRHOCORIDAE, p. 212

Ocelli present.

TL generally >13 mm; scent glands present.

TL generally <10 mm; scent glands absent.
RHOPALIDAE, p. 206

Head as wider or wider than pronotum.
ALYDIDAE, p. 206

Head narrower than pronotum.
COREIDAE, p. 206

185

Further, the authors of different keys may use different terminology for the same anatomical structure. This is especially true when keys have been created decades or more apart. Stay tuned, for better news is ahead.

How do keys work? In a book or online, they present the user with worded couplets describing a particular physical character on the specimen. These couplets may or may not be accompanied by illustrated examples. Some keys are accompanied by SEMs (scanning electron micrographs) because don't we all have an electron microscope in the garage? If a character on your specimen is lacking, that may be one option in the couplet, leading you to the next couplet, and so forth, until you eventually arrive at a genus identification or a full scientific name, including the species. Such traditional keys were once found solely in scientific journals, monographs revising various genera, and in technical books.

Today, there are online interactive keys (matrix keys), richly illustrated, that can add and subtract various specimen attributes in one fell swoop. The more modern field guides may have flow charts that act similarly to keys, but it can be tricky following the arrows from one figure to another, especially across several pages.

The only method for success in advanced insect identification is practice, practice, practice. Look at *both sides* of your bilaterally symmetrical specimen. The author once mistakenly keyed a wasp specimen to a genus found only on another continent. It turned out one leg had a missing spine, while its complement on the opposite side had all spines intact.

Lastly, take heart in knowing that even expert specialists are not always able to reach a conclusive result. Dissection of genitalia is often required for

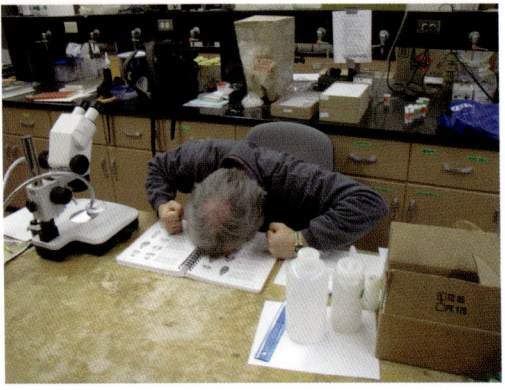

The ecstasy and the agony: Successfully navigating a key to the precise identification of a specimen is exceptionally gratifying. Sadly, pure frustration is too often the outcome instead. (George Boettner)

species identification. Increasingly, cryptic species are being discovered based on DNA analysis. Even then, some species can be essentially identical genetically, while exhibiting starkly different behaviors when alive. Several fireflies can only be identified by their flash patterns, for example.

— QUALITATIVE PROJECTS

THERE EXIST MANY opportunities, run by scientists, for assisting in more controlled and detailed surveys and other endeavors. In the Grand Canyon, for example, there are Monarch butterfly tagging studies, bumble bee diversity and population censuses, firefly tracking, and aquatic insect monitoring, to name but a few. One project involves catching female solitary wasps (*Cerceris fumipennis*) carrying beetle prey, to see if they are catching the Emerald Ash Borer. There are even opportunities to work alongside scientists abroad, through organizations like Earthwatch.

Field surveys: Many agencies and organizations rely on volunteers to help with surveys of butterflies, bees, and other insects, tracking their abundance and diversity over time. This includes meticulous recording of observations and strict adherence to transects (delimited physical areas) and time limits. (KAREN POVEY)

— DONATIONS

FINANCIAL DONATIONS AND in-kind donations of equipment, books, and other materials are usually welcomed by scientists and their institutions. This is especially true for professionals in countries where such resources are scarce. Idea Wild is one such organization that collects equipment for dispersal to Indigenous scientists in need around the globe. Closer to home, libraries, local chapters of the Audubon Society, and related educational and conservation groups may have programs in place to accept binoculars and other equipment to use in their programs.

— VOLUNTEERING

THE MAJORITY OF natural history museums, insect zoos, and butterfly houses in the United States rely on docents to do all manner of tasks. These can include pinning specimens collected by scientists, interpreting insect diversity and behavior for museum or zoo guests, and leading tours. Such work can sometimes lead to paid employment at the facility. Elementary and secondary schools and organizations such as 4-H and Scouting America also welcome volunteer mentors to teach effective techniques to the next generation of bugwatchers.

— ADDITIONAL IDEAS

ADVANCED BUGWATCHING MAY look like you sharing your own observations and discoveries with others, in person and through exhibits. Approach garden clubs, nature centers, libraries, and civic organizations, and offer to do a presentation at one of their meetings or events. While community science platforms are the place to deposit your observations and associated data, presentations to others are where you get to tell *stories*. Meanwhile, ask if you can print out your photos and pin them in library displays to help promote books about insects.

— CALLING ALL ARTISTS

THERE IS A place for you in entomology and science communication. Scientific illustrators contribute to publications that describe new species, revise genera, and otherwise add to the scientific record. See the Guild of Natural Science Illustrators website for more about this unique career. Less formal works created by painters, cartoonists, glass artists, sculptors, and fabric artists help the public learn to love insects in a fun way.

— *PATRICK HSIEH (HE/HIM)*

HI! MY PATH to observing insects has been a long and winding one that has led me to a unique brand of bugwatching: I hunt for fossil insects, when I am not working as an environmental scientist with the State of California.

As a child, I was afraid of bugs. They were dirty and creepy. I had to confront that fear in high school freshman biology class, when I was required to complete an insect collection. That meant catching, touching (shudder), and pinning specimens (serious eek). What I did not expect was that this exercise would foster a deeper relationship

Patrick Hsieh (KELLI HSIEH)

with my father. He grew up in a rural setting, and his childhood was spent hunting and catching insects as playmates. My high school biology assignment was the perfect excuse for him to pass on his long-dormant skills and passions.

Today, with a child of my own, my perspective has evolved. I want my child to get out into nature, and I foster an appreciation of the natural world, as opposed to the virtual one, for the sake of their mental health and myriad other reasons.

My interest in palaeoentomology presents unique challenges. The first step is to research where insect fossils might be found. Access to known fossil beds can be difficult, as many are on private land. Luckily, those on public land are somewhat easier to get onto. Still, public agency policies change constantly, and you sometimes have to advocate for entry. Mobilizing your peers is vital to advancing the science.

It is critically important to take good, detailed notes on where and how fossils are found. I specialize in compression fossils, which are understudied compared to amber inclusions. Once collected, the specimens are prepared for study. Then I label them and image them with my digital microscope. The intention is that my collection will be properly documented and conserved, and be of value to future researchers.

When observing fossil insects, there is always the possibility that you are the first person to make a significant find. One personal thrill was identifying the oldest (to my knowledge) stalk-eyed fly, in the family Diopsidae. The males of these flies have eyes mounted on stalks as long, or longer, than their bodies. It was immensely satisfying to identify one and to send it to a researcher who was working on them.

My advice to those new to bugwatching is to try to network with others who have similar interests. Here in Southern California, there is the Lorquin Entomological Society. Even if there is not a club that is easily accessible to where you live, many clubs have hybrid virtual/in-person meetings and would be happy to have you attend. Look for the nearest club you can find.

CHAPTER 9

THE FUTURE OF BUGWATCHING

Projecting over the horizon, bugwatching looks both bleak and promising. The bad news is that there are fewer bugs to watch than there used to be. It might not be the "insect apocalypse," but there is general agreement that insect abundance and diversity are declining. With this in mind, we cannot have enough bugwatchers filling in the knowledge chasms and effecting positive change in invertebrate conservation. How will advances in technology enhance our ability to assess insect population levels, and make and share observations? Are we willing to expand the universe of bugwatchers to include all stakeholders, regardless of their gender identity, race, ethnicity, level of affluence, ability or disability, social or solitary preference, age, or level of education?

Endangered species: A team of scientists and students releases American Burying Beetles, *Nicrophorus americanus*, into the wild. This is a highly endangered species, now bred in captivity at a handful of zoos. (Mandy Pritchard for Cincinnati Zoo & Botanical Garden and Kelli Walker, respectively)

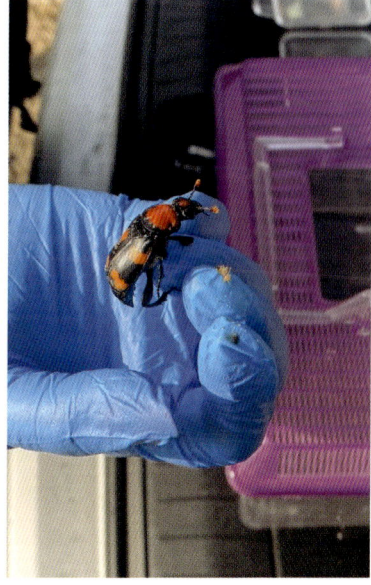

— REASONS FOR INSECT DECLINE

CLIMATE CHANGE IS the mega-problem threatening biodiversity as a whole, but the other factors impacting insects, and wildlife in general, are the same as they have always been. It pays to briefly revisit those threats, then discuss how we can address them.

— Habitat destruction and fragmentation.
— Growing footprint of industrial agriculture.
— Continued reliance on pesticides.
— Urbanization, urban sprawl, plus associated road mortality.
— Pollution, including light and noise pollution.
— Deregulated global commerce.
— Excessive consumerism leading to more resource extraction.

The stressors listed above are themselves contributors to the climate crisis. Habitat is destroyed to make way for large-scale agriculture, both crops and livestock, and for urban housing, business, and recreation. At the grotesque scale of many agricultural enterprises, the only way they can thwart pests is through the application of insecticides and other chemicals. The soil becomes exhausted, necessitating fertilizers that, like the other chemicals, contaminate water through runoff into rivers and lakes, and percolation into underground aquifers. What is not plowed under is paved over for other needs, mostly corporate in nature. The extent of our global economy magnifies the carbon footprint of humanity through transportation of goods and people to all corners of the world. We all participate in consumerism, but this is disproportionately a Western cultural behavior in its degree of excess.

It is worth noting that species cease to properly function in ecosystems long before they become extinct. Sheer abundance today does not guarantee that extinction cannot come tomorrow. Look at the Passenger Pigeon and the Rocky Mountain Locust as cautionary tales.

— WHAT YOU CAN DO

START BY UNDERSTANDING how economic interests conspire to create expectations in the marketplace and standards of personal behavior. Agricultural methods are used to design and maintain urban and suburban "lawnscapes," for instance, and the requisite fertilizer and pesticide treatments. We need to recognize our personal role in subscribing to these destructive practices and alter our behavior accordingly. Exercise your vote, your voice, and your choices in the marketplace, where and when you can.

Conserving biodiversity requires sacrifices. Okay, maybe not to *that* degree, but there is something to be said for getting out of your comfort zone. (ERIC R. EATON)

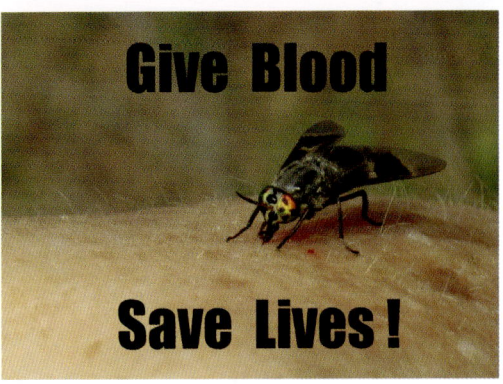

Give Blood

Save Lives !

Switch off outdoor lighting at night, or install less harmful designs in fixtures. Try motion-sensor lights. See if your city or town has a dark sky initiative. Dark Sky International is the premier organization for resources to address light pollution at the local level and beyond.

Decrease consumption of corporate retail products. Patronize local agriculture. Shop at farmers markets. Participate in community gardens and urban farms. Look for co-ops instead of chain stores. Grow your own vegetables and

When you have the only native prairie or meadow in your subdivision, you get all the butterflies. Enticing your neighbors to follow suit is more of a challenge. (BENJAMIN VOGT)

other foods, and share with neighbors, homeless shelters, and soup kitchens, as local law permits.

Landscape with native plants as much as you can tolerate or indulge. You may be fighting city weed ordinances and HOA (homeowners association) regulations, but precedents are being set with increasing frequency for how to successfully challenge those.

Please do not keep honey bees. Honey bees, genus *Apis*, are social bees occurring naturally in Europe, Asia, and parts of Africa. Elsewhere, they are a heavily managed commodity, essentially the slave labor of pollination for industrial agriculture.

— ABOUT INVASIVE SPECIES

THE TERM *INVASIVE* species has come under attack recently. It implies that any species described as foreign should be met with a militaristic response and an intent to eradicate. Understandably, human immigrants may compare this mentality to how *they* are depicted: as innately offensive, probably criminally so. Further, tolerance or acceptance of novel flora and fauna hinges almost entirely on economic impact. If a species is perceived as a threat to human health, agriculture, forestry, or other industry, then we show no mercy. We need to recognize our own complicity. If we covet organisms and products from overseas, we are responsible for any troublesome outcomes. It is not the fault of the organism itself for succeeding in a new homeland.

That said, there are instances where a displaced species has seriously compromised the ecological integrity of indigenous (presettlement) habitats and landscapes. The arrival and spread of the Emerald Ash Borer triggered preemptive cutting of ash trees in North America, depriving at least 98 native ash-dependent insect species of their host trees. It is much more difficult to argue against active intervention in cases like that. Going forward, caution should be the new standard, ahead of profit for businesses that traffic in ornamental plants, for example.

— EXPANDING THE REACH AND IMPACT OF BUGWATCHING

IDEALLY, BUGWATCHING SHOULD be as great a pursuit as birding. Indeed, we can take lessons from the birding community to design and construct a global movement in that direction. We can also better anticipate potential pitfalls by learning how birding has negotiated the potholes. Education, engagement, advocacy, and diversity, equity, and inclusion should all play a part in how that happens. Only then can invertebrate conservation efforts succeed.

When will bugwatching become as big an industry as birding? At the Biggest Week in American Birding, avian enthusiasts line a boardwalk at Magee Marsh in northwestern Ohio. (ERIC R. EATON)

— EDUCATION

OVERCOMING ENTOMOPHOBIA IS a challenge. Insects remain alien to the majority of people, creatures so far removed from us that they fail to evoke a sense of kinship, sympathy, or any other positive sentiment. This view reinforces an artificial species hierarchy that harmfully distances ourselves from arthropods.

Consumers are conditioned by advertisements and the media to view most insects as pests that require a product or a service to control or eradicate them. You know better. Show people that they can save money by avoiding the unnecessary purchase of over-the-counter pesticides or contracting for repeat visits from pest control services. Emphasize simple measures that prevent insects from becoming a problem in the first place, and demonstrate that most other insects can be tolerated with little difficulty.

Few people are aware of the *benefits* provided by insects, especially ecosystem services like pollination, seed dispersal, and the decomposition of decaying organic matter, as well as the place of insects in the food web. Fewer still understand that insects are threatened, endangered, and in need of conservation. These should be constant points of emphasis for both professional entomologists and enthusiastic bugwatchers, through public engagement.

— ENGAGEMENT

It can be argued that engaging young children is the best and fastest way to promote a lifelong appreciation of the natural world, but teaching adults can effect immediate positive change. Maybe we can follow the lead of Japan, where there is a tradition of keeping insects (*mushi*), especially singing insects and beetles, as pets. We could at least use Pokémon as a springboard; we could progress from searching for virtual fictional creatures to searching for even stranger real-life insects. Be a mentor that way. Buy children butterfly nets, magnifiers, and other gear. They can catch and release or make a permanent collection. Let them decide.

Be an influencer. Your enthusiasm is contagious, and your patience in helping people learn is limitless. Share accurate, educational posts on your social

Be a mentor: The author and his first mentor, Jim Anderson, sit on a bench to look over an insect collection at Sunriver Resort in eastern Oregon. Such nurturing relationships can last a lifetime, to the benefit of both parties. (Sue Anderson)

media accounts, or create your own. Make better memes. Most memes are overly simplified, with no sources indicated for further investigation by the audience. Enroll your artistically inclined friends in your campaign. Post images of insects you find, with a brief explanation of what makes the species "cool" and valuable.

— *ADVOCACY*

MOST CONSERVATION ORGANIZATIONS are constantly battling against "development" of wildlife habitat. This needs to be a bigger concern and activity for bugwatchers, too. Your favorite spots could be bulldozed for housing projects; existing parks could be encircled by subdivisions and business or shopping centers. Speak out against such projects, if it makes sense, and especially if you have data to back up your assertions. Maybe the only population of a dragonfly species in the entire state resides at that location, for example. It may fall on deaf ears, but try anyway. You may recruit more people to bugwatching and conservation as a result.

We should encourage parks agencies, outdoor outfitters, and tourism providers to give equal weight to "passive" nature recreation, instead of continuing

Backpacks to lend: Audubon of Kansas provides libraries with "Adventurepacks" filled with nature-study equipment, for lending to those who cannot afford to purchase their own gear. We need more initiatives like this. (DR. JACKIE AUGUSTINE)

to prioritize recreation that is "nature conquering." Mountain biking, rock climbing, and trail running are examples of destination-oriented pursuits that can be destructive to natural resources.

Natural areas, parks, and attractions need to be accessible via public transportation. Transit currently focuses on getting people to work, shopping centers, and park-and-ride hubs. This needs to change in the interest of public health alone.

— DIVERSITY AND INCLUSION

BESIDES THE SUGGESTIONS offered in chapter 7 on Social Bugwatching, there are additional ways to expand the community of bugwatchers and drive progressive initiatives:

— RECOGNIZE INDIGENOUS EXPERTS as equal partners and custodians of biodiversity knowledge. Give them the power to decide what information they share with the group or organization.

Butterfly smiles: Dr. Jana Johnson and her team celebrate the release of captive-bred Palos Verdes Blue butterflies by giving the American Sign Language sign for "butterfly," with one maverick "moth" sign at left. (DR. JANA JOHNSON)

— IN THE NAME of human inclusion, endorse efforts to correct common English names of organisms that currently reflect animosity toward various ethnic groups or that are named after individuals known to be slave owners, racists, misogynists, and bigots.
— ENCOURAGE AND SUPPORT Indigenous initiatives for Rights of Nature legislation. Ecuador, Bolivia, Panama, New Zealand, and India have recognized rights of nature at the national level. In the United States, it has been largely at the level of towns and sometimes cities, like Pittsburgh, Pennsylvania.

— WHAT NOT TO DO

CERTAIN STRATEGIES AND practices in the conservation of insects cannot continue. Further, the influence of the pest control industry is out of control,

Languishing in obscurity, the Palos Verdes Blue butterfly, *Glaucopsyche lygdamus palosverdensis*, is far more endangered than the Monarch, but does not benefit from the same degree of publicity. (MAX A. SPRUTE)

No Mow May is a convenient and catchy slogan, but the tactic does not apply everywhere. The best solutions for preserving and enhancing insect diversity are more complex and labor intensive. (ANNE READEL)

and technology is both a blessing and a curse. Bugwatchers must play a role in driving necessary change.

There is a belief that a single-species focus—such as on particular butterflies, like the Monarch—is the way to protect adjacent species and the habitats they dwell in. This results in a disproportionate amount of dollars and personnel that could better serve an array of species, or preserve habitat through acquisition, conservation easements, and other methods, or rehabilitate and restore habitat of marginal quality.

Similarly, butterflies are imagined as "flagship species" that provide a gateway to an appreciation of other insects. This is rarely effective. Anyone engaged in public outreach hears questions such as "How do I get rid of the other insects eating the milkweed I planted for *my* Monarchs?"! Honey bee propaganda is worse, in any arena other than agribusiness, where these insects are necessary for crop pollination at the unsustainable scale that currently exists.

Slogans, quick fixes, and fads are meme-able, but problematic. "No Mow May" and "Leave the Leaves" are catchy phrases, but the reality is more nuanced. Often, such campaigns are rolled out prematurely, before enough studies prove their effectiveness. People receive conflicting advice, and end up doing nothing because they cannot determine the best course of action.

Artificial Intelligence (AI) is a mixed bag in relation to bugwatching. *Pattern-recognition AI* is likely to improve algorithms used to identify insects, interpret their behaviors, and predict trends in abundance and diversity. On the other hand, *generative AI*, such as Dall-E and ChatGPT, already generates images of insect "species" that do not exist, among other forms of misinformation that is all too easily widely broadcast at the expense of real science.

— *SKYE AUSTIN (THEY/THEM)*

Skye Austin (SKYE AUSTIN)

Hello! I've been a bugwatcher since I was a small child. We had this giant butterfly bush in our backyard that was buzzing with insect diversity—bees, butterflies, flies, spiders, you name it! It was mesmerizing to watch, especially when the evening light hit it just right. I've maintained my love of bugs throughout my K–12 schooling and into university, where I learned that you can make a career of studying them! As an undergraduate, I was quickly picked out as the bug person, and I loved getting messages, texts, or people coming up in person to ask, "Skye, what is this thing!?"

One way I bugwatch is to pick a spot and sit for a while; also, I wander around vegetated areas. Method one can be particularly useful in urban areas; I used this method on my college campus, as I didn't have a car until the end of my senior year, so transportation to natural sites was difficult. I've had great success sitting in the grass under trees. This approach is especially helpful on higher-pain days, when I want to go outside but wandering around much isn't feasible. Method two is better for large expanses of greenspace, like a park, a riparian buffer, a meadow, or a garden. On a quick stroll, you'll see larger insects—bumble bees, butterflies, and some larger beetles. If you take it slow and look closely at vegetation and other landscape elements, you'll notice smaller organisms like leafhoppers and smaller wasps, bees, and ants.

The main challenge I face is accessibility—I use a combination of forearm crutches and a wheelchair to get around, and many natural areas are far from wheelchair accessible. I think it's also worth noting that some of the more rural areas of the United States are outright dangerous for people of color and visibly queer folk.

Using crutches in the field can be tricky, especially if I'm carrying equipment, but I've gotten the hang of it. As people work to lower barriers to the field of ecology, accessibility of field sites has become a topic of conversation; however, there's a lot of work to be done. As the conversation evolves, I hope that positive changes can occur to make ecology and field biology accessible to as many people as possible.

Seeing a bug I haven't seen before is so amazing! I see fun new bugs all the time; I love bending down and saying, "Oh, hello! Who are you?"—much to the confusion of anyone around me. Watching bugs crawl or fly around, minding their business, also brings a smile to my face, especially if they land on me. Bugwatching moments that have made me feel proud are being able to identify something from its gestalt or purely off of "vibes," and holding large moths. There's something so goofy about them—they're such a joy!

A FINAL WORD

THE AUTHOR CAN trace his own fascination with insects and related animals to the influence of his kindergarten teacher. She was a talented artist and had drawn a trapdoor spider on the blackboard one day. Her explanation of its behavior was captivating. As an only child, I felt disconnected from my peers. I couldn't easily stand up for myself, but I could learn about spiders, bugs, bats, snakes, and sharks, and tell the other kids why *those* creatures were cool. It was a kinship with the underdog that has only gotten stronger.

What kind of social lens should we be looking through in our relationships with other species and each other? Do we have a proper respect and reverence for all the players? Are we willing to make ourselves vulnerable in order to attain higher collective goals? These are the kinds of deep and relevant questions we need to be asking. We will never succeed at conserving and protecting biodiversity if we do not respect and accept *human* diversity.

The power of narrative cannot be overemphasized. Facts are great, but storytelling brings those facts to life and complements the scientific with the sacred and the personal. It is through stories that we best connect with one another and develop respect and empathy.

Hang in there, like these gall midges suspended on a spider thread. The more time you devote to the pleasures of bugwatching, the more you will be rewarded. (KIMBERLY R. FLEMING)

Hang in there!

We can do better at facilitating accessibility for all. Why can't we have field guides in languages other than English, and as audiobooks? Widespread, affordable broadband service is needed to bridge the digital divide.

Conservation organizations should continue enlarging their spheres of emphasis to include insects and related arthropods. They must also actively expand their comfort zone of human demographics. We need all hands, all eyes, indeed all senses, and personal strengths, directed toward common goals. It is the only way forward if we seek to grow public understanding of the natural world and accelerate positive, lasting impacts on the ecosphere. Think big, think "bug."

It is the author's hope that this book has given the reader food for thought, sparked fond memories, answered unspoken questions, and inspired a brighter vision of stewardship for the planet. Go forth, now. Walk slowly, look closely, and be curious.

RECOMMENDED RESOURCES

— BOOKS

Brown, Tom, Jr. 1987. *Tom Brown's Field Guide to the Forgotten Wilderness*. New York: Berkley Books. 222 pp.

Cavalier, Darlene, Catherine Hoffman, and Caren Cooper. 2020. *The Field Guide to Citizen Science: How You Can Contribute to Scientific Research and Make a Difference*. Portland, Oregon: Timber Press. 187 pp.

Dethier, Vincent G. 1992. *Crickets and Katydids, Concerts and Solos*. Cambridge, Massachusetts: Harvard University Press. 140 pp.

Evans, Howard E. 1968. *Life on a Little-known Planet*. New York: E. P. Dutton. 318 pp.

Hatch, Warren A. 2020. *In One Yard: Close to Nature Book 2*. Portland, Oregon: Warren Hatch. 215 pp.

Preston-Mafham, Rod, and Ken Preston-Mafham. 1993. *The Encyclopedia of Land Invertebrate Behaviour*. Cambridge, Massachusetts: The MIT Press. 320 pp.

Russell, Helen Ross. 1971. *Winter Search Party: A Guide to Insects and Other Invertebrates*. Nashville, Tennessee: Thomas Nelson. 171 pp.

Stokes, Donald W. 1983. *A Guide to Observing Insect Lives*. Boston: Little, Brown. 371 pp.

Teale, Edwin Way. 1937. *Grassroot Jungles*. New York: Dodd, Mead. 240 pp.

Voisard, Lisa. 2024. *Insectorama: The Marvelous World of Insects*. Lausanne, Switzerland: Helvetiq. 224 pp.

— FIELD GUIDES

Abbott, John C., and Kendra Abbott. 2023. *Insects of North America*. Princeton, New Jersey: Princeton University Press. 585 pp.

Eaton, Eric R., and Kenn Kaufman. 2007. *Kaufman Field Guide to Insects of North America*. Boston: Houghton Mifflin. 392 pp.

Eiseman, Charley, and Noah Charney. 2010. *Tracks & Sign of Insects and Other Invertebrates: A Guide to North American Species*. Mechanicsburg, Pennsylvania: Stackpole Books. 582 pp.

Evans, Arthur V. 2008. *National Wildlife Federation Field Guide to Insects and Spiders of North America*. New York: Sterling. 497 pp.

Peterson, Merrill A. 2018. *Pacific Northwest Insects*. Seattle: Seattle Audubon Society. 520 pp.

Will, Kipp, Joyce Gross, Daniel Rubinoff, and Jerry A. Powell. 2020. *Field Guide to California Insects* (2nd edition). Oakland: University of California Press. 521 pp.

— ACTIVITIES AND PROJECTS

Allison, Linda. 1988. *The Wild Inside: Sierra Club's Guide to the Great Indoors*. Boston: Little, Brown. 144 pp.

Brenner, Kelly. 2023. *The Naturalist at Home: Projects for Discovering the Hidden World Around Us*. Seattle, Washington: Mountaineers Books. 222 pp.

Brown, Vinson. 1968. *How to Follow the Adventures of Insects*. Boston: Little, Brown. 201 pp.

Headstrom, Richard. 1963. *Adventures with Insects*. New York: Dover. 221 pp.

Imes, Rick. 1992. *The Practical Entomologist*. London: Aurum Press. 160 pp.

Lawrence, Gale. 1986. *The Indoor Naturalist; Observing the World of Nature Inside Your Home*. New York: Simon & Schuster. 210 pp.

Nardi, James B. 1988. *Close Encounters with Insects and Spiders*. Ames: Iowa State University Press. 185 pp.

Tallamy, Douglas W. 2019. *Nature's Best Hope: A New Approach to Conservation That Starts in Your Yard*. Portland, Oregon: Timber Press. 254 pp.

Vogt, Benjamin. 2017. *A New Garden Ethic: Cultivating Defiant Compassion for an Uncertain Future*. Gabriola Island, British Columbia: New Society. 179 pp.

— ONLINE

— *COMMUNITY SCIENCE AND CONSERVATION*

Bugguide: https://bugguide.net/node/view/15740

iNaturalist: https://www.inaturalist.org/

Moth Photographers Group: https://mothphotographersgroup.msstate.edu/

Odonata Central (dragonflies & damselflies): https://www.odonatacentral.org/#/

Singing Insects of North America: https://orthsoc.org/sina/i00abt.htm

Songs of Insects: A Guide to the Voices of Crickets, Katydids & Cicadas: https://songsofinsects.com/

Xerces Society Community Science: https://xerces.org/community-science

— *DIVERSITY, EQUITY, AND INCLUSION IN NATURAL SCIENCES*

Birdability: https://www.birdability.org/

Entomologists of Color: https://www.entopoc.org/

Field Inclusive, Inc.: https://www.fieldinclusive.org/

Hearts for Sight Foundation: https://heartsforsightfoundation.org/

Idea Wild: https://ideawild.org/

Nature for All (Los Angeles, California): https://lanatureforall.org/

Out in the Field (OITF), an initiative of the Wildlife Society: https://wildlife.org/out-in-the-field/

Outdoor Afro: https://outdoorafro.org/

Outdoors for Everyone: https://www.outdoorsforeveryone.org/

The Venture Out Project: https://www.ventureoutproject.com/

Seeing the Forest for the Queers: https://foreststewardsguild.org/seeing-the-forest-for-the-queers/

— *BLOGS*

Arizona: Beetles, Bugs, Birds, and More: https://arizonabeetlesbugsbirdsandmore.
   blogspot.com/
Beetles in the Bush: https://beetlesinthebush.com/
Bug Eric: https://bugeric.blogspot.com/
Bug Tracks: https://bugtracks.wordpress.com/
Entomology Today: https://entomologytoday.org/
MObugs: https://mobugs.blogspot.com/
Ohio Birds and Biodiversity: https://jimmccormac.blogspot.com/
The Bug Chicks: https://www.thebugchicks.com/articles
The Smaller Majority: https://thesmallermajority.com/

— *PODCASTS*

Arthro-Pod: https://arthro-pod.blogspot.com/
Bug Banter (Xerces Society): https://www.xerces.org/bug-banter
Can I Bug You?: https://www.youtube.com/watch?v=e6VDXS2zUY4
Nature's Archive: https://naturesarchive.com/
Ologies: https://www.alieward.com/ologies

— *YOUTUBE*

Ant Lab: https://www.youtube.com/@AntLab
Chris Kline: https://www.youtube.com/@chriskline6488
Chromatophone Nature: https://www.youtube.com/c/
   ChromatophoneProductionsMedia
Distractible By Nature: https://www.youtube.com/@distractiblebynature
Eric Eaton: https://www.youtube.com/channel/UCE-
   QgIMUOT5cOzOeDmQZlZw
Insectopia: https://www.youtube.com/@Insectopia
KQED Deep Look: https://www.youtube.com/@KQEDDeepLook
SciBugs: https://www.youtube.com/@SciBugs

ACKNOWLEDGMENTS

Many thanks to Robert Kirk and the editorial board at Princeton University Press for agreeing to publish this book. I am honored and flattered by the endorsement. Thank you, Samantha Gallagher, for lending your artistic talents to this book, bringing to life the diversity of insects and their astonishing behaviors. Literary agent Russell Galen suggested improvements to my proposal for this book, without which the project may not have seen the light of day. Gregory S. Paulson read over the first draft and helped improve the manuscript, and Patricia Fogarty's copy editing that made the work stronger still. Megan Mendonça was incredibly generous with her time in guiding me through the photo permissions process. I wish to recognize again the late, great Steven J. Prchal, who coined the term *bugwatching* and founded the Sonoran Arthropod Studies Institute (SASI) in Tucson, Arizona. He also initiated the Invertebrates in Education and Conservation Conference that still takes place in southern Arizona. He was a profound influence, offering support and opportunities to me at the beginning of my career, as he did for countless others. Erin Starkey's energy and fascination continue to fuel my passion and dedication to bugwatching, even after she succumbed to depression at such a young age.

Special thanks to Dr. Lisa Rainsong, Chris Kline, Shelly Cox, Charley Eiseman, Joseph Saunders, Patrick Hsieh, Krystle Hickman, Skye Austin, and Bryna B. for volunteering to share their experiences by writing profiles for this book. Thank you, Arlo Chavez, for sending a PDF of a reference I sorely needed. The late Dr. Elizabeth Bernays shared with me her essay "In Praise of Looking" shortly before she passed away. Lila Higgins and Sam Tayag at the Natural History Museum of Los Angeles County initiated our conversation about inclusion and accessibility, freely sharing their own experiences and ideas.

Many friends, mentors, colleagues, and organizations offered to share photos from their archives, including Anne Readel, Karen Povey, Audrey Sauble, Kathy Carroll, Kim Moore, Mike Quinn, Ryan Bridge, Nancy McIntyre, Nancy Miorelli, and Rob Holquist.

Without the wholehearted support of my partner, Heidi Eaton, this work would not have been possible. She has more patience with me, and faith *in* me, than I have for myself, and keeps me flourishing in every way. Thank you, sweetheart!

INDEX